'In a fluent and readable style of writing, this doctor critically exposes the link between our genetically determined ageing process and our diet. This is highly recommended for anyone who consciously wants to eat more healthily and is, given the emphasis on prevention, especially useful for policy makers.'

DR HERMAN BECQ, internist-endocrinologist

'Nutrition is an important factor in not only the prevention, but also the onset, of cancer. This book makes the reader examine our socially integrated habits and really think about what we eat.' DR AN VANDENBROECK, cancer specialist

'The author's greatest gift is that he provides a deeper insight into universal nutritional problems and in particular, those in western society. He uses plain language to bring a banal and media-driven topic into clear view.'

DR HANS DECOSTER, cardiologist

'A healthy diet is the best medicine, as this book shows. If the recommendations of Verburgh were followed by everyone, healthcare would become both a lot more fun and cheaper. A must-read!'

ADJIEDJ BAKAS, Trend-watcher of the Year and author of *The Future of Health*

'Governments and health authorities that worry about trends like the worldwide ageing population and completely un-affordable healthcare would do well to read The Food Hour-glass: the simple yet rigorously justified solution lies in a healthier diet and lifestyle!'

REINIER EVERS, Trend watcher of the Year, founder of Trendwatching.com

D0047482

THE
FOOD HOURGLASS

Stay younger for longer + lose weight

DR KRIS VERBURGH

This book contains ideas and opinions of the author. The knowledge obtained from this book is not intended to treat or cure particular illnesses or medical conditions. Neither the author nor the publisher wishes this book to be used as a means to provide professional medical services, healthcare services or other services. Before making any changes to your lifestyle and eating habits, always consult a physician or other professional healthcare providers. The author and publisher cannot be held responsible for any loss or risk, either personal or non-personal, which is sustained by applying, either directly or indirectly, content from this book.

HarperThorsons
An imprint of HarperCollins*Publishers*
77–85 Fulham Palace Road,
Hammersmith, London W6 8JB
www.harpercollins.co.uk

and *HarperThorsons* are trademarks of
HarperCollins*Publishers*

First published in Holland as *De Voedselzandloper* by
Prometheus Books 2012.
First published in the UK by HarperCollins*Publishers* 2014

3 5 7 9 10 8 6 4 2

© Kris Verburgh 2014

Kris Verburgh asserts the moral right to
be identified as the author of this work

A catalogue record of this book is
available from the British Library

ISBN 978-0-00-755616-8

Printed and bound in Great Britain by Clays Ltd, St Ives plc

All rights reserved. No part of this publication may be reproduced, stored in a retrieval system, or transmitted, in any form or by any means, electronic, mechanical, photocopying, recording or otherwise, without the prior written permission of the publishers.

MIX
Paper from
responsible sources
FSC C007454

FSC is a non-profit international organisation established to promote the responsible management of the world's forests. Products carrying the FSC label are independently certified to assure consumers that they come from forests that are managed to meet the social, economic and ecological needs of present and future generations.

Find out more about HarperCollins and the environment at
www.harpercollins.co.uk/green

Contents

Introduction

Actually, I had never planned on writing a 'diet book'. Truth be told, the last thing I ever wanted to do as a doctor and scientist was write a book about dieting. Diet and weight loss weren't exactly top-ranking issues that occupied my brain. What's more, I had somewhat of an aversion towards all those diet books and self-proclaimed health gurus who kept coming up with the strangest of diets.

What had been occupying my mind for many years was the ageing process. How and why we get older is an immensely fascinating process from a philosophical, evolutionary and biochemical viewpoint. Already as an adolescent I would devour books and scientific articles about ageing. One of the first things I learnt was the vital importance of nutrition in the ageing process. The rate at which we age is greatly determined by what and how we eat.

For example, research has shown that sugar metabolism plays an important role in ageing. Sugar not only causes wrinkles, cataracts or arterial stiffness through AGEs (Advanced Glycation End-products, as we'll explain later), it also plays a direct role in our life expectancy. Scientists are able to triple the life expectancy of worms by genetically modifying the genes that play a role in sugar and insulin metabolism.[1-2] If these worms are then also put on a special diet, they can live six times longer. Other research studies reveal that rats fed on a calorie-restricted diet, but who still receive optimal nourishment, can reach the ripe old age of 1800 days. In human terms, that's 150 years old! Above all, these animals have far fewer age-associated illnesses such as cancer, heart disease or dementia. Yet more research shows that protein-rich diets cause laboratory animals to age considerably faster and die younger. And so on, and so forth.

Professor Michael Rose, an authority in the field of ageing (and whose famous experiments could double the age of fruit flies), says the following about nutrition and ageing:

> Nematode, fruit fly, rodent, and clinical findings [in humans] all implicate metabolism in the control of ageing. Food turns up in all of these organisms as a major modulator of ageing. [. . .] The use of food energy affects the rate of ageing.

So you see, this is no ordinary diet book. It takes into account the view that nutrition plays a crucially important role in the ageing process. This is of course logical, when you consider that nutrition is the engine of our metabolism and that metabolism drives all our bodily processes. With this knowledge in mind, I have put together a diet (I don't really like the word 'diet', and will come back to this later). In contrast to most diets, the primary aim of this diet is not to lose weight. The – often considerable – weight loss is merely a fortunate side-effect. The purpose of this diet is first and foremost to slow down the ageing process. It seeks to make sure that we grow older more slowly and that we are less likely to suffer the illnesses associated with old age. Nearly all popular diets are based on the principle of quickly losing weight, but this is the wrong attitude. A genuinely good diet should always put priority on remaining healthy as long as possible. The weight loss then follows automatically. My background in biogerontology (the science that researches ageing) has provided me with a great deal of knowledge about metabolism and the ageing process, allowing me to see through most diets and have a rather accurate estimate of their harmful and unhealthy long-term impacts.

What I discovered during my research into the how and why of ageing is that most diseases that plague western society are essentially ageing-associated: heart disease, dementia, osteoporosis, type-2 diabetes, deterioration of hearing and eyesight, muscular atrophy and increased body fat percentage, high blood pressure, cataracts, etc. All these ageing-associated diseases could be combated at the same time by conducting research into the ageing process itself, instead of just looking at certain age-related illnesses like diabetes or dementia.

The chance of contracting many of these age-associated diseases can be dramatically reduced with a healthy diet. I've remoulded this healthy diet into a new and easy-to-use model: 'the food hourglass'. This food hourglass can be used as an alternative to today's food pyramid and food plate, which we are now seeing everywhere and which, according to academic research, are out-dated (and actually always have been).

My dietary pattern looks at adding quality to life, as well as combating age-associated diseases. Someone with a healthy dietary pattern not only reduces the risk of chronic illnesses in the future, but also feels the direct benefits, such as more energy, better concentration, a more positive state of mind, more drive to be active, etc. Also, quite a few medical ailments like heartburn, an irritable bowel, fatigue or high blood pressure are all too often the result of an unhealthy diet.

In short, the food hourglass doesn't merely look at adding more years to your life, but also more life to your years.

SUMMARY

Most diets focus on **weight loss**, something that should never be the primary goal of a good diet.

A good diet aims at maintaining **health** and slowing down the **ageing process**. Weight loss then follows automatically.

The food hourglass is a new alternative to the food pyramid and the food plate.

The food hourglass strives for healthy weight loss and a slowing down of the ageing process.

A Note for Doctors and Dieticians

I refer to numerous studies in this book. Some of these studies show, for example, that people who eat a handful of walnuts every day lower the risk of a heart attack by 45 percent[a], or that those who drink green tea on a regular basis have 21 percent less chance of suffering a stroke.[b] I mention these studies because they reveal an important tip of the iceberg: namely that healthy food, like walnuts or drinking a cup of green tea really can have a significant effect on our health, especially when this is part of an all-round healthy diet. Despite this, there are some people who don't believe that nutrition can have a large impact on our health.

Luckily there are actually different kinds of studies that even more convincingly demonstrate the power of healthy nutrition. While the walnut and the green tea studies used questionnaires to assess dietary habits, there are also *intervention studies*. Here, researchers carry out actual interventions: they deliberately alter the eating habits of their test subjects, and then wait to see what happens. In this way, researchers from the University of Newcastle succeeded in completely reversing type-2 diabetes in just eight weeks. The patients taking part in this study were put on a (strict) diet, which cut out bread, potatoes, pasta and rice, but did include plenty of vegetables. In eight weeks, their blood sugar levels were normal, the fatty degeneration of the liver was five times less and the pancreas (which produces insulin) was working normally, and this is all without medication.[c]

Or take heart disease. Previously, it was impossible to

claim that the plaque (the 'chalky' mass on the artery walls which silts up the blood vessels) could be reduced. Recent research has shown that the silting up of the blood vessels can be reversed, and this can be achieved by diet. Researchers from Harvard University assembled a number of patients who were on the waiting list for a heart operation. By putting these people on a special diet, almost 80 percent of the patients no longer needed an operation. They could be crossed off the operating list. Not only that, but due to this new approach, those people on the diet had 10 times less chance of suffering a heart attack than those patients who were having the normal treatment.[d]

These kinds of studies are important, because they show that 'chronic' illnesses can actually be reversed. The important point here is that the 'diets' or changes in the nutritional habits have to be sufficiently thorough. The nutritional recommendations offered by the government, hospitals and official bodies often don't go far enough. You cannot reverse diabetes by simply replacing white bread with the wholemeal variety. However, if you include eating less bread, potatoes, rice and pasta for a while in your advice to a diabetes patient, then you can bring about significant improvements. It's hardly surprising that doctors are sometimes less than enthusiastic about diets: patients that follow the official government diet recommendations rarely achieve significant improvements (the average reduction of the HbA1c – a measure for the saccharification ('glycation') of the red blood cells – is only 0.4 percent for patients who follow an official 'diabetes diet'). These meagre results can be blamed on the fact that government recommendations are for various reasons not very effective and could be much healthier.[e] Many studies have shown this convincingly. Heart patients who followed an unofficial, more Mediterranean-inspired diet (plenty of vegetables, fruit, good quality oils, nuts, ...) had 70 percent less chance of dying during the study than patients who followed the official low-fat diet recommended by the American Heart Association, the organisation in the USA which issues all kinds of health guidelines.[f] After 2.5 years, the study was cancelled because it was deemed 'unethical' to allow patients to carry on with the official diet recommended by the American Heart Association. Another study in which diabetes patients followed a more vegetarian diet, saw a three-times greater improvement in their blood sugar levels than those who followed the official diet of the American Diabetes Association, the diet recommended for diabetes patients by almost all hospitals.[g]

Even these non-standard diets, which can lead to so many improvements, could be better, as I have tried to show in this book.

It is these along with many other studies that have encouraged universities such as Harvard, illustrious hospitals such as the Mayo Clinics and countries like Austria to create completely different nutritional models. The basis of the Austrian food pyramid and that of the Mayo Clinic for example, are no longer based on bread, potatoes, pasta and rice, but on vegetables and fruit.

Almost as a last resort, the proponents of the official nutrition guidelines often claim that the other diets are too difficult for patients to follow, or require major changes. The beauty of it is that patients do actually stick to these diets more readily because they soon see the health benefits and can reduce their medication, and because, ironically enough, these diets are often intrinsically simpler than the official diet guidelines with their long lists of 'forbidden foods' or obscure recommendations.

I could cite many other intervention studies, from heavily uncontrolled diabetes patients who, thanks to oatmeal, are able to use 40 percent less insulin,[h] to researchers from Oxford University who gave older patients B-vitamins and discovered that the brains of their patients showed 7 times less shrinkage as they grew older.[i] You might call these figures nit-picking and believe that it is better not to tell patients about this kind of thing because they just might live on nothing but oatmeal or swallow pounds of B-vitamins. I really believe that we should tell our patients about these studies, so that they can see how important nutrition is to our health. They have the right to know that good nutrition can prevent or slow down the onset of certain illnesses associated with ageing, and sometimes even reverse the effects. Health professionals who want to know more about the scientific background of the food hourglass, can read the 'For health professionals' section at the end of the book. Please also note that for every study mentioned in this book, dozens more can be found in the medical databases.

a. Hu, F. B. & Stampfer, M. J. Nut consumption and risk of coronary heart disease: a review of epidemiologic evidence. *Curr. Atheroscler. Rep.* 1, 204–9 (1999).

b. Green and black tea consumption and risk of stroke: a meta-analysis. *Stroke.* 40, 1786–92 (2009).

c. Lim, E. L. *et al.* Reversal of type-2 diabetes: normalisation of beta cell function in association with decreased pancreas and liver triacylglycerol. *Diabetologia* 54, 2506–14 (2011).

d. Pischke, C. R. *et al.* Clinical events in coronary heart disease patients with an ejection fraction of 40 percent or less: 3-year follow-up results. *J. Cardiovasc. Nurs.* 25, E8–E15

e. Willett, W. C. & Stampfer, M. J. Rebuilding the food pyramid. *Sci. Am.* 288, 64–71 (2003).

f. De Lorgeril, M. *et al.* Mediterranean alpha-linolenic acid-rich diet in secondary prevention of coronary heart disease. *Lancet* 343, 1454–9 (1994).

g. Barnard, N. D. *et al.* A low-fat vegan diet improves glycaemic control and cardiovascular risk factors in a randomized clinical trial in individuals with type-2 diabetes. *Diabetes Care* 29, 1777–83 (2006).

h. Lammert, A. *et al.* Clinical benefit of a short-term dietary oatmeal intervention in patients with type-2 diabetes and severe insulin resistance: a pilot study. *Exp. Clin. Endocrinol. Diabetes* 116, 132–4 (2008).

i. Douaud, G. *et al.* Preventing Alzheimer's disease-related gray matter atrophy by B-vitamin treatment. *Proc. Natl. Acad. Sci. U. S. A.* 110, 9523–8 (2013).

1

About medicine, diets and nutrition

We place a huge amount of faith in medicine. Maybe even too much faith. We believe that when we're feeling ill, the doctor and a whole team of nurses and paramedics are ready and able to cure us. They are definitely ready, but modern medicine – despite its thousands of new medications, expensive scanners and ultra-modern operation rooms – is not able to heal most illnesses, whether it's a common cold, or a heart attack, a stroke, osteoporosis, lower back pain due to arthritis, dementia or a disease of the nervous system such as multiple sclerosis. By and large, only antibiotics, chemotherapeutics and some surgical operations can truly cure people of their disease. There is no actual treatment that can cure a heart attack, a stroke or dementia. Even the common cold or bronchitis can't be cured. In over 90 percent of cases a virus causes a cold or bronchitis, and the only reason we recover from our cold or bronchitis is because our immune system beats the virus. Painkillers like paracetamol and aspirin do nothing more than suppress the symptoms of a cold. They reduce the pain by suppressing the immune system; therefore enabling the sickness to actually last longer.

As a young medical student, I often came across patients entering the doctor's surgery with hopeful expectations of having their ailments cured only to find out, usually much later, that their specific condition actually cannot be cured. And then they were left to get on with their life with Crohn's disease, rheumatism, nervous system disorders or a failing heart. It quickly made me realise that people put far too much faith in medicine. While, ironically, they place far too little faith in their own capacity to prevent diseases and to maintain their health levels. And it's our dietary pattern that plays the most important role in achieving this.

It goes without saying that almost everyone knows that nutrition is important for our health. We're bombarded daily with nutritional advice in magazines, on TV and by health experts, all constantly reminding us that too much sugar or an excessive fat intake are unhealthy for us. What's more, we've all had our own mothers or grandmothers telling us to eat our greens because they're so healthy. But what none of us ever had was a mother or a health expert telling us exactly how healthy vegetables really are and how unhealthy other foods are and just how important the influence of food is for our health, the ageing process and our longevity. Let me give you just a few examples.

In Japan, where (for the time being) there is still a dietary pattern that differs from that in western culture, prostate cancer occurs ten times less often than in the West.[3] This enormous difference has nothing to do with the genetic differences between Japanese and Caucasian men: Japanese men who emigrate to the United States and who follow more western dietary habits, have an equal chance of prostate cancer as the average American; so ten times more than in their original homeland. Prostate cancer is the most common form of cancer among men: as long as a man gets old enough, he's almost sure to get it. At least 30 percent of men in the age group of 70 to 79 years old have a prostate tumour (whether discovered or not). There isn't a single preventive medicine on the market today that can reduce the chance of getting prostate cancer, let alone by a factor of ten. However, a healthy diet can.

Cancer, one of the most important causes of death in the West, is a lot less common in certain regions of Asia. Here

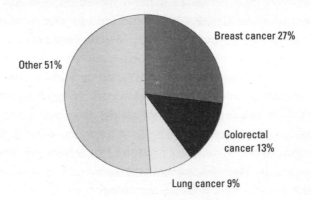

Cancer in Western women

Breast cancer 27%

Other 51%

Colorectal cancer 13%

Lung cancer 9%

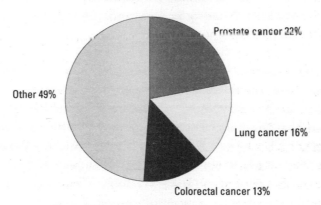

Cancer in Western men

Prostate cancer 22%

Other 49%

Lung cancer 16%

Colorectal cancer 13%

Prostate cancer and lung cancer are the two most prevailing forms of cancer in men in western society. Two lifestyle changes (eating more healthily and smoking less) can contribute to a significant reduction to the number of cancer patients. *Source: globocan 2008, World Health Organisation*

we're talking about orders of magnitude of five to ten times less. Medically speaking these are huge differences because doctors usually get very excited about any substance that reduces the growth of cancer by five to ten percent. Nutrition plays a prominent role in the risk of getting cancer. If you don't smoke, then about 40 percent of the risk of cancer is determined by your eating habits. This is not mere conjecture, but, the conclusion of the World Cancer Research Fund (WCRF) report (2007). This report contains the findings of thousands of research studies; it took over five years to compile and includes the recommendations of ten world-renowned experts in collaboration with 234 oncologists and scientists – many of their findings are discussed in this book. So, if you don't smoke, your diet determines 40 percent of the risk of getting cancer. The remaining 60 percent of the risk factor is determined by a number and variety of reasons, such as genetic predisposition, infectious diseases, sun exposure, pollution, sexual habits, certain drugs and medical procedures, a sedentary lifestyle, particular professions, and so on.

So how can nutrition reduce the risk of cancer? Several research studies show that substances in foods, such as broccoli and herbs, contain strong protective agents against cancer. For instance, a certain agent in parsley can slow down the blood vessel formation around tumours, just as the medicine Gleevec can. Gleevec is a new kind of anticancer drug that's made in laboratories. Scientists describe Gleevec as a 'wonder drug' against cancer that slows the growth of both blood vessel cells around tumours and of tumours themselves. In the field of blood vessel growth, parsley was, however, eons ahead of the scientists. This is not surprising, since many of our most powerful medications come from plants. Aspirin, the well-known anti-inflammatory drug and painkiller, is derived from a compound from the willow bark. Metformin, the most prescribed anti-diabetic medicine in the world, stems from the French lilac. And powerful chemotherapeutic drugs, used in hospitals, like taxanes and vinco alkaloids come from the yew tree and the periwinkle plant.

*   *   *

Let's now have a look at age-related macular degeneration, a medical condition whereby the retinal cells in the eye die off due to oxidative and other damage, which results in an irrevocable decline of vision. Macular degeneration is one of the major causes of blindness in western society (together with diabetes, another 'accelerated ageing disease' that is making a steady climb in countries with an unhealthy dietary pattern). At least twenty percent of people over the age of 60 have macular degeneration, but we can state that every older person is in some way affected by the damaging processes that eventually cause macular degeneration. Ophthalmologists (eye doctors) have no real effective treatment for this medical condition. However, research that was published in the prestigious medical journal *The Journal of the American Medical Association* (*JAMA*) reveals that people who eat vegetables on a daily basis have fifty percent less chance of developing this eye disease.[4] If this is then combined with other healthy foods, for instance by eating fatty fish, the chance of macular degeneration is reduced even further. A large study in which 681 twins took part, and that appeared in the journal of ophthalmologists – *Archives of Ophthalmology*, showed that those who eat enough fatty fish have 45 percent less chance of developing macular degeneration.[5]

The scarcity of prostate cancer in Japan, powerful anticancer substances in herbs such as parsley, and the importance of eating vegetables and fatty fish in preventing macular degeneration are just a few of many important insights that partially lift the veil, and which explain that, with the right diet, a whole array of chronic illnesses could be delayed or even prevented, diseases that modern medicine has the utmost difficulty in treating after diagnosis (when it's actually too late). We should in fact have less faith in medicine and have more faith in a healthy lifestyle.

*   *   *

There will come a time when historians will look back at this era. If they were to name the two most important medical discoveries of our age then, according to the French-American doctor David Servan-Schreiber, they would have to mention the following two events:

1 the discovery of antibiotics;
2 the enormous impact of a healthy lifestyle on the prevention of diseases.

Numerous scientists and trend watchers believe that in the next decades the governments, health insurance companies and an increasing number of doctors and patients will see the importance of good nutrition in the prevention of chronic diseases. Insurance companies will concentrate more and more on nutrition and preventive medicine to reduce the enormous costs inflicted by chronic diseases. They may choose to increase the premiums for people who smoke, have an unhealthy diet or don't have a gym subscription.

This won't simply be a way to get people to pay more for their insurance, it will be a pure necessity because health costs for chronic diseases will continue to rise until, at a certain point, they will become unaffordable.

A scenario like this is already starting to unfold in the United States. The US spends the most per capita on healthcare, which is double that of most other western countries – more than 7,000 dollars per inhabitant per year, while most European countries spend approximately 3,000 dollars a year per inhabitant. In 2011, the US spent 2,700 billion dollars on healthcare, equivalent to nearly 20 percent of the total government expenditure. In 2019, this will increase to approximately 4,600 billion dollars. In spite of this gigantic expenditure, the US still finds itself somewhere in the middle of the table when it comes to health indicators, among which, life expectancy is one. In short, pumping more money into healthcare does not lead to a healthier population. The American government is starting to realise this and a huge number of initiatives are being put into place to reduce medical costs and improve the general health of the population at the same time. An important pillar is preventive medicine with an emphasis on the importance of a healthy diet in the prevention of chronic diseases.

What is a healthy diet?

So, honestly, what is a healthy diet? Every day we're exposed to a tsunami of health advice through popular magazines, talk shows, documentaries and cooking programmes. And we haven't even touched on the internet and the torrents of health and diet books. These books have often been written by self-proclaimed 'experts' who promote theirs as the only true method. Usually this method is incredibly easy to follow (after all, the general public should be able to follow it daily). For instance, avoiding sugars and eating a lot of protein – a blow in the face to every biochemist who knows that the metabolism cannot be caught out by an Atkins diet or a high-protein diet. Every doctor or scientist knows that it is extremely difficult to outsmart nature.

In addition, these diet gurus often justify their own diet with countless pseudo-scientific arguments, or even worse, with insights from actual published scientific studies. As we will see in a moment, it is important to understand that a great deal of scientific research is not conducted correctly or is interpreted wrongly. This explains the contradictory health advice, spread via the media, which leads to people no longer knowing what's healthy and what's not.

But it is still possible to see the wood for the trees and to understand what a healthy lifestyle really means. This is possible by relying on well-conducted, large-scale studies that appear in prominent scientific journals and by calling upon the sound knowledge of biochemistry and the workings of the human body as well as by considering insights and findings derived from other disciplines, such as evolutionary biology and the science of the ageing process. In this book, I will try to inform you about this as much as possible by showing you that there are many food and health misconceptions circulating. Among other things, I'd like to prove to you the following points:

- that most fats play no role in cardiovascular diseases;
- that dairy products are not as healthy as we always assumed;
- that most antioxidants don't work and can be dangerous in the long term;
- that today's food pyramid and food plate are out-dated;
- that resveratrol, the famous anti-ageing substance in red wine, does not help us to live longer;
- that omega-3 fats are less healthy than fish oil;
- that exercising with the aim to lose weight is not a good idea;
- that most diets are damaging to your health;
- that products such as green tea are not healthy because of the antioxidants;
- that the only scientifically proven method that considerably slows down the ageing process is not a medication or a 'super antioxidant pill', but rather a certain dietary habit.

Some of these insights may prove to be quite a revelation for some people. For instance, we could ask ourselves the question why so many people believe that antioxidants or dairy products are healthy. There are two major reasons. Firstly, the people or industries that spread such messages often have a hidden agenda (such as an anti-ageing guru wanting to write a bestseller, or the dairy industry promoting the sale of milk). They then actually attempt to substantiate their message through scientific research studies that show the health benefits of antioxidants or dairy products. What a lot of people don't realise is that most scientific studies aren't conducted properly (for example, because they have too few test subjects or, because the test subjects are not properly divided or, because no comparison is made with a placebo (a fake medicine) or with a control group). Also, scientific studies are often incorrectly interpreted by popular media. This helps health gurus and the food industry to always find scientific research that supports their agenda.

That is precisely why it is so important that any conclusions from one study are re-examined and reproduced by other, more experienced research groups. Only after the same conclusions have been made by several research groups and these conclusions are published in leading scientific journals can certain assumptions be adopted. For instance, let's look at the report by the World Cancer Research Fund (WCRF). For this report 500,000 research studies about cancer and nutrition were collected. Poor quality studies were disregarded. This left just 22,000 research studies remaining that were up to standard! Of those studies, only 7,000 that met the strict criteria of the WCRF were ultimately selected. In short, only 1.4 percent of the research studies about the causes of cancer were of a high enough quality to serve as valid insights concerning the origin and prevention of cancer. However, most authors of diet books and health books base their theories on smaller, abundantly available and poorly conducted research studies that, for example, show substance X protects you from cancer or heart disease, without providing evidence from large-scale research studies or 'meta-analyses'. A health guru can easily rely on thousands of poorly conducted research studies and then proclaim that drinking alkaline water daily, or taking high doses of vitamin C, can drastically reduce the chance of developing cancer (we'll be talking about the myths and science about vitamin C later on).

Now, how can we find out which research studies are executed properly and which are done poorly? It's important to know that there are different types of medical and scientific journals – those with a high and those with a low 'impact factor', a measurement for the reputation of the journal. Research studies that appear in journals with a high impact factor are more reliable, because they have been properly executed, often use a large number of test subjects and are conducted by researchers with extensive experience. The two scientific journals with the highest impact factor are *Nature* and *Science*. If your scientific work is published in one of these two journals, you have made it as an academic and universities around the world will be rolling out the red carpet in an attempt to employ you as a researcher. Medical journals that have a very high impact factor are: *The Lancet, New England Journal of Medicine* or *The Journal of the American Medical Association*. Every medical specialism has its own authoritative journals with a high impact factor, such as *Circulation* (cardiovascular diseases), *Gastroenterology* (stomach and intestinal diseases), *Archives of General Psychiatry*, etc. Studies that are published in such journals have been more rigorously conducted and are more reliable than the vast majority of research studies that appear in journals with a low impact factor. Most studies that are cited in health and diet books have appeared in such journals.

Journals with a low impact factor far exceed the number of those with a high impact factor and of course, it is much easier to get published in one of these. This is not a problem in itself: as a researcher you have to start somewhere. Generally, these types of journals make sure that substance X is put in the spotlight in the popular media. However, in the meantime, that same substance is being tested by different scientists in the academic world with a better research setup, more test subjects, improved statistical methods and analyses, and so forth. This often means that years (or even decades) go by, only for it to emerge that substance X has no influence whatsoever on heart disease or cancer.

A well-known example is vitamin E. Some years ago several research studies (usually appearing in magazines with a low impact factor) stated that vitamin E reduces the risk of cardiovascular disease. But larger and better research studies since, published in renowned journals, show that vitamin E does not help against cardiovascular diseases. Yet to this day, a vast number of health gurus still preach in their books that vitamin E is such a wonderful antioxidant: a super antioxidant that, whether or not combined with vitamin C, can dramatically reduce the chance of developing heart disease. These gurus will undoubtedly cite a whole array of studies to prove their point. But this is not how science works.

Another example is the coenzyme Q10. On the web and in a great deal of magazines, you cannot avoid the articles proclaiming that the coenzyme Q10 can slow down ageing. The research studies referred to have usually appeared in journals with a low impact factor. For example, research carried out by the scientist Emile Bliznakov showed that 17-month-old mice administered with Q10 lived on average an extra 11 months, while mice that did not receive the coenzyme Q10 only lived an additional 5 months:[6] a doubling of the remaining life expectancy! This research resulted in a whole host of health books that hailed the coenzyme Q10 as a kind of elixir of life; music to the ears of coenzyme Q10 manufacturers. The study by Bliznakov appeared in 1981, but it was only in the 1990s that other scientists made the effort to conduct further research. Mice were stuffed with coenzyme Q10 in high and low doses and in many different antioxidant cocktails. And guess what? Coenzyme Q10 did not extend the lifespan of mice. Dr Lonnrot, one of the coenzyme Q10 researchers, believes that the mice used by Bliznakov had a coenzyme Q10 deficiency. These are mice that cannot produce enough Coenzyme Q10 themselves. It's no wonder that by administering coenzyme Q10, Bliznakov's mice were able to live longer. I should mention however, that some major research studies have shown Q10 in very large doses can slow down Parkinson's disease. But the message remains the same: coenzyme Q10 has no influence whatsoever on the longevity of healthy mice and people. The moral of this story: before ingesting a handful of pills of a particular substance because research says mice live longer because of it, it is important to wait for more reliable studies to appear in journals with a high impact factor.

Another reason why we're constantly bombarded with sub-standard health advice is because medical research studies are often incorrectly interpreted by popular media. Imagine a certain research study that shows substance X lowers cholesterol. This study is then picked up by a few journalists, swiftly read through and without further delay published in newspapers, weekly magazines and health websites, proclaiming that 'substance X fights cardiovascular diseases', an incorrect conclusion.

This false conclusion is reached because of the following rationale: substance X lowers cholesterol. That's what the research showed. It is common knowledge that excess cholesterol causes cardiovascular diseases so it seems logical to most people that if substance X lowers cholesterol and an excessive level of cholesterol causes heart disease, substance X must prevent cardiovascular diseases. But medically speaking, this is a fallacy. Just because substance X lowers the cholesterol level, this doesn't mean it automatically has a positive effect on the cardiovascular system. Besides lowering the cholesterol level, substance X may also increase the level of other substances that are actually bad for the heart and blood vessels; substances that were not researched during the study. Each substance can affect thousands of kinds of enzymes and cellular mechanisms, all of which can either have a positive or negative influence on heart and vascular diseases. For this reason you should not jump to the conclusion that X is healthy for heart and blood vessels just because it lowers the cholesterol level if the research study does not clearly show that substance X actually decreases the chance of a heart attack.

Examples of substance X are medicines called 'fibrates'. Fibrates are used in order to lower the level of cholesterol and fats in the bloodstream. It's absolutely true that fibrates drastically reduce cholesterol and fats in the bloodstream. At first glance this is positive for the heart and blood vessels. But fibrates also increase levels of homocysteine in the blood, which is a risk factor for cardiovascular disease. Research has shown that fibrates do not reduce the chance of death due to a heart attack, even though they can spectacularly reduce the levels of cholesterol and fat levels in the blood. In a similar manner, health gurus can place almost any substance in a positive light. They can even claim that smoking is healthy, given the fact that research has indeed shown that it reduces the chance of Parkinson's disease. But of course they conveniently forget to mention the fact that smoking not only affects the brain, but also not so very unimportant organs like the lungs.

Another fallacy that often occurs when interpreting scientific and medical research studies is the following. If a research study demonstrates that a shortage of substance Y causes certain health problems, some people will argue that, by taking a large amount of substance Y, these health issues will be preventable. But that's not how the human body works. For instance, a selenium deficiency is known to be a cause of cancer, yet if too much selenium is taken (which is easily done because it's relatively toxic) it can actually cause cancer, as well as a range of other health issues. And the same applies to antioxidants and vitamins.

In short, if I want to pose as a health guru, all I have to do is write a book about how vitamin E is healthy for the heart and blood vessels and how the antioxidant enzyme Q10 can extend your life. I can even back this up with references to many kinds of different scientific research studies that support the theory, while avoiding other studies that show there is no effect at all. Health gurus often get away with the above-mentioned practices because they make excellent use of one of today's larger problems: an oversupply of information.

A real diet is not a diet

Anyone who is on a diet is going about it the wrong way. After all, dieting assumes that you will make the extra effort to eat less during a certain period. This is completely ridiculous. Why on earth would you try to lose weight during a certain period, when you know that afterwards, that lost weight will return as soon as you begin to eat 'normally' again?

A genuinely good and healthy diet is one that never ends. It's a lifestyle change that needs to be learnt and maintained for the rest of your life so that the long-term effects on health can truly be of benefit. What's more, losing weight should never be an effort, because if it is it can never be maintained. It is also wrong to assume that you have to eat less to lose weight. If your diet is healthy, then you are allowed to eat as much as you like: if you are overweight, you'll lose weight regardless. The only good diet is a dietary pattern that turns into a lifestyle. However, because humans are animals of habit, it can be difficult at first to adjust to and maintain a new and life-changing dietary pattern. Therefore, extra motivation to stick to this dietary pattern comes from the knowledge that it not only helps you lose weight, but it also slows down the ageing process.

Fortunately, the human body is extremely capable of maintaining new habits, so when the old habits have finally been broken and the new habits have been learnt, these healthy habits will then almost always be anchored in daily life forever. After a few months, a newly-learnt dietary pattern will already no longer feel like a new way of life. What's more, other things happen to the body; healthier nutrition causes other taste sensations to develop; grapes and strawberries suddenly become wonderfully tasty sweets. Feeling hungry will be a new sensation; it will no longer be a pseudo-hungry feeling that is primarily a feeling of weakness and a lack of concentration coupled with a strong craving for sugars brought on because your blood-sugar level has plummeted since your last sugar rush; it will be a 'real hunger' accompanied by an urge to eat healthy food.

I have moulded my dietary pattern, a nutritional lifestyle that I would like to introduce in this book, into a model which I call the 'food hourglass', an analogy based on the food pyramid and the food plate. The dietary pattern that I describe here is a combination of insights from both medical and scientific literature. I haven't discovered the ultimate health diet, but I am continuing to build on the 'shoulders of giants'; shoulders of reputable doctors and researchers who recommend certain nutritional patterns because the benefits have been documented in medical literature. Some of these 'diets' are for example the CRON-diet (Calorie Restriction under Optimal Nutrition) by Dr Walford, the Dr Fuhrman diet or the Japanese Okinawa diet. While these nutritional guides are all extremely healthy, they can be made healthier still or more comfortable. For instance, the CRON-diet is very healthy in that it can significantly increase life expectancy but it is difficult for many people to sustain. The Dr Fuhrman diet concentrates too heavily on being vegetarian. Vegetarians live longer on average but their maximum lifespan is not increased (more on this later). What all of these nutritional patterns have in common with each other is that they have been substantiated by scientific arguments. When we use these scientific insights as the benchmark only a select few truly healthy diets are left over from the sumptuous offerings of existing diets.

Why most diets are unhealthy

It has already been stated that we are constantly being bombarded with new diets and conflicting nutritional advice. Coffee is healthy according to one magazine, yet to the next it causes arrhythmia. Expert A claims that fatty fish is healthy, while expert B says fatty fish is full of mercury. One study states that green tea causes bladder cancer, while another study believes green tea actually protects against cancer. According to diet X we can eat a lot of sugars, as long as we eat a low-fat diet, and diet Y tells us that fats are fine and that sugars are the real culprits. The list is endless.

I can already give you one tip: never ever get your dietary advice from the popular media! Some nutritional advice and diets are downright absurd, like the 'blood type diet', whereby you eat according to your specific blood type. This is a joke even though there are already millions of books on the bookshelves. Many diets that seem to make a little more sense are also not very healthy. Take the world-famous classic Atkins diet. Dr Atkins recommended eating as few sugars (carbohydrates) as possible and as many protein-rich and fatty foods as possible. He provides a very extensive medical explanation as to why this is important, and some-where in his story he is correct: too much sugar is very unhealthy. But, and I can't stress this enough, you have to look at the whole picture. Every medical student learns that an excess of proteins is also unhealthy, not least of all because metabolising the proteins is difficult for the liver and kidneys. It is important to know that from about our thirtieth year, the function of our kidneys deteriorates by 10 percent per decade, and that this process is accelerated by an excessive intake of proteins. When patients go to hospital with liver or kidney failure, they are put on a strict low-protein diet. Moreover, as we will see, there are other more important reasons why a high protein intake is not healthy in the long term.

Atkins does have a point in saying that sugars are unhealthy. But even on this point he is still wrong in his approach. Allow me to explain. Too much sugar is unhealthy, partly because they make us age more quickly. Sugar molecules that are in the body are sticky; they cause proteins to stick together more easily. This makes our tissues more rigid and makes them lose their elasticity. This is one reason we get wrinkles and why, as we get older our blood pressure increases (due to the walls of the blood vessels becoming harder). At first glance, Atkins takes the right approach by eliminating sugars as much as possible. But when someone has a very low sugar intake, as the Atkins diet prescribes, the body goes into 'ketosis'. Ketosis is a state whereby the body makes large amounts of 'sugar' itself. But this is no ordinary sugar, this is a 'super sugar', called methylglyoxal. Methylglyoxal is 40,000 times more reactive than normal sugar and makes proteins stick together like a mad Pritt stick. Research shows that people following the Atkins diet have double the amount of methylglyoxal in their body, the same amount 'is only found in poorly controlled diabetics'. Dr Aubrey de Grey, a famous specialist in the field of ageing, calls the Atkins diet a 'potential recipe for accelerated ageing, not a reprieve from it'.[7]

But how was it possible that the Atkins diet could become so popular? I'll give you a couple of reasons:

1 The Atkins diet does in fact make sure you lose weight. It works and that is the most important thing for a lot of people.
2 The diet cleverly responds to what people want to believe: that you can lose weight and still continue to eat fat and protein-rich food.
3 The diet has a few scientific arguments that sound very plausible to those people who don't know the full biochemical picture.
4 It is simply part of human nature that allows hypes to quickly happen (and pass).

Besides the Atkins diet there is a whole host of other popular diets, such as the pasta diet or the Beverly Hills diet. These diets are such an amalgamation of ignorance, marketing and pseudo-scientific drivel that it´s not even worth me wasting any ink explaining.

SUMMARY

Many **age-associated diseases**, such as prostate cancer, cardio-vascular diseases or macular degeneration can be prevented or delayed by a healthy diet.

In the future, the **costs of chronic diseases** will become unaffordable for our healthcare system. Programmes aimed at prevention, with the emphasis in healthy food, can reduce these costs dramatically.

There is a great deal of **contradictory dietary advice** because:
– a great deal of medical studies are too small-scale or are not executed correctly, for instance due to a poor partition of test

subjects, no use of fake medicines (placebos) as a comparison, poor statistical analysis, etc;
- a great deal of medical studies that appear in popular media are naïvely interpreted or simply misinterpreted.

Most **diets** are unhealthy and **health gurus** can substantiate their claims with scientific studies because:
- they base their insights on small-scale or incorrectly executed studies published in scientific magazines with a low impact factor;
- they make far-reaching conclusions (not every cholesterol-lowering agent automatically protects against heart attacks);
- they cite only those studies that confirm their claims and ignore research studies that refute their arguments;
- they concentrate on one biochemical mechanism and ignore the complete biochemical picture (replacing 'unhealthy' sugars with 'healthy' proteins according to the Atkins diet);
- they focus primarily on weight loss and do not take into account the long-term consequences, such as the shortening of life expectancy. A diet or dietary pattern from the viewpoint of biogerontology, the science of the ageing process, takes this into account.

2
Pyramids, plates and hourglasses

You probably know them: the food pyramid or the food plate, which tell us what we should eat. The food pyramid is used in most European countries. Recently, in the US the food pyramid was replaced by a plate. In Great Britain the food plate rules.

The food pyramid usually looks like this:

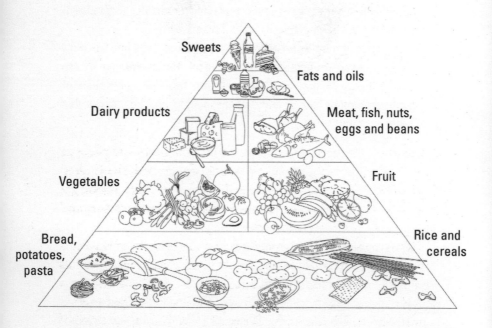

The base of the food pyramid consists of starchy products (bread, pasta, rice and potatoes). The section above contains vegetables and fruit, in approximately the same amounts. Above this you'll find foods such as meat and dairy products. In the top area are fats and oils, as well as sweets (sugars and other candy).

To a large extent, the food plate resembles the food pyramid regarding the content:

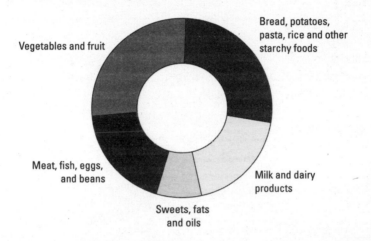

Vegetables and fruit take up an equal portion of the plate as the starchy foods (bread, potatoes, pasta and rice); then meats, fish, and eggs are collected. Milk and dairy foods get a prominent place. Fats should be taken in moderation and are given a small section within the plate.

With regard to the shape, I am not a supporter of the food pyramid and even less of the food plate. At least the pyramid has a clear overview. When I look at the food plate I see a heap of food that has been thrown together. The pyramid is constructed with hierarchical steps so that the importance of certain food groups is at least accentuated by placing them either above or below the other.

But of greater importance is the content of the pyramid or

plate and according to scientific research, this content is out-dated. Yet today we still find both pictures in canteens, company restaurants, in countless health magazines and on websites.

Harvard University has not let this go unnoticed and their department of nutrition has come up with an alternative food pyramid. The objective of their pyramid is to offer a better and healthier alternative to the standard pyramid and plate.

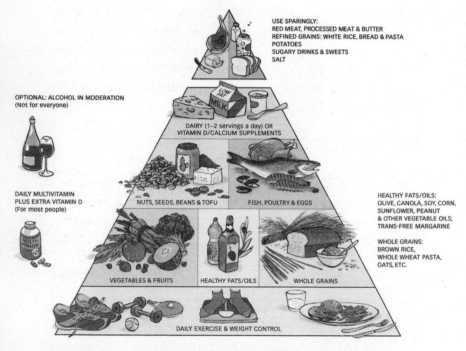

© 2008 Harvard University. For more information on The Healthy Eating Pyramid, go to The Nutrition Source, Department of Nutrition, Harvard School of Public Health, www.thenutritionsource. org, and Eat, Drink, and Be Healthy by Walter C. Willett M.D. and Patrick J. Skerrett (2005), Free Press/Simon & Schuster Inc.

What stands out? The base of the pyramid is different: instead of only grain products, the base is now shared with vegetables, fruit and ... fats (which can also be healthy). What is also remarkable is that potatoes, white rice, white bread and non-whole wheat pasta have been moved all the way to the top: they're now in the 'forbidden' top section of the pyramid and have been pushed into the same corner as crisps, soft drinks and sugar waffles. Red meat has also been moved to the forbidden top section. The unhealthy fats remain at the top; the healthy fats are at the bottom. Newly added is a recommended intake of dietary supplements. This seems strange – aren't there enough vitamins in your food when you follow a varied and healthy diet? We will touch on this later. Another interesting point is that the researchers at Harvard consider dairy products to be relatively unhealthy.

Food plate and food pyramid	Harvard Healthy Eating Pyramid
There is no emphasis on the dangers of potatoes, white rice, white bread and non-whole wheat pasta.	Potatoes are no longer found in the base, but rather in the 'forbidden' top, along with white rice, white bread and non-whole wheat pasta, alongside more notable unhealthy foods such as sweets and soft drinks.
Starchy products (including bread, pasta, rice) form the base of the food pyramid and the food plate.	Starchy products, together with vegetables, fruit and healthy oils, form the base.
Dairy products (including cheese, yoghurt, milk) are recommended.	Dairy products may be fully replaced by dietary supplements with vitamin D and calcium.

Table continued overleaf

Food plate and food pyramid	Harvard Healthy Eating Pyramid
Dietary supplements are not discussed.	Dietary supplements are recommended.
Red meat is recommended.	Red meat is discouraged and placed in the 'forbidden' top. Fish and white meat such as chicken are recommended.
Fats and oils are unhealthy.	Healthy fats and oils belong to the base.
Little attention is paid to nuts, beans and tofu.	Nuts, beans and tofu serve as important meat substitutes.
No attention is given to alcoholic beverages.	Moderate use of alcohol is recommended.

Today's food pyramid and food plate are outdated and you are probably asking yourself why the food pyramid and the food plate are still promoted and recommended. To me, as someone who grew up in the world of medicine, this is no surprise. By the time new concepts establish themselves, many years (or decades) have gone by. Insights from the scientific world usually trickle into our society at a woefully slow pace.

Above all, there are some powerful interests involved in deciding what you put on your plate every day. One of the reasons the old food pyramid and food plate can still be found everywhere you look is because the grain, meat and dairy industries use their lobbying power with governments to keep the pyramid and the plate in their current form: a form with a strong focus on grains, meats and dairy products. It's therefore no coincidence that these foods take up a large portion of the pyramid and the plate. The grain, meat and dairy industries are not only the guardians of the pyramid and the plate, they also had a great influence on their creation. Most European food pyramids and plates are based on the classic American food pyramid. The agriculture industry in the United States has, for many decades, had a considerable influence on large government agencies. To a large extent, it determined what the American food pyramid should look like; it is not without reason that it was developed by the US Department of Agriculture and not by the Department of Health. If it had been created by the Department of Health, then it probably would have looked a lot healthier and grain, meat and dairy products would not have received such prominent positions. You could argue that the current pyramid and plate primarily serve the health of the agricultural sector and not the health of the general public.

Lobbying by powerful industries is not something to be underestimated within the health sector and the scientific world. Take Meir Stampfer for instance, a renowned professor of nutritional sciences at Harvard University. He has published some well-known studies on the harmful influence of sugars on the heart and blood vessels. In an interview in *Scientific American*, he makes it very clear how 'the soft drink industry is lobbying very hard to bring his studies into disrepute'.[8] What right do soft drink manufacturers have to interfere with scientific research, let alone to bring into doubt reputable professors and medical studies that can benefit the health of millions of people?

Of course there are even more reasons why the old food pyramid and food plate remain intact. Government institutions that provide nutritional information often tend to have a condescending view of the general public (something that really annoys me as a doctor). As far as they're concerned, changes to the food pyramid and food plate should be kept to a minimum, because too many changes would be too difficult for the general public to understand and adopt. Robert Post, a senior official within the US government and co-director of the Center for Nutrition Policy and Promotion, was asked whether he would inform the public of some of the important new insights about fats and sugars. He said he would not because messages to the public should be kept 'short and simple' and 'to the point'.[8] However: our body is not 'simple' in its workings. Plus: advice concerning our health should not be 'short'. First and foremost, it should be sound. In short, government agencies are jeopardising our health because they believe that some scientific insights are too complex for the general public to understand.

But we also come across posters of the food pyramid and food plate in a lot of hospitals. Don't the doctors know that these concepts are outdated? To be honest, most doctors don't concern themselves much with nutrition. Not much time is devoted to nutrition at all during medical studies. Doctors are mainly educated to deal with acute medicine; the main priority of their study focuses on treating illnesses when in fact it's already too late, when a tumour breaks through after twenty years of growth, or when an artery, after its been silting up for thirty years, is finally fully clogged, with the inevitable heart attack that follows – a process which is markedly affected by our dietary pattern. Little attention is given to preventive medicine during the years of medical study for doctors.

Professor Doctor Richard Smith from the authoritative *British Medical Journal* says the following about nutrition and doctors:

> *Many doctors' knowledge of nutrition is rudimentary. Most feel much more comfortable with drugs than foods, and the 'food as medicine' philosophy of Hippocrates has been largely neglected. That may be about to change. Concern about obesity is rocketing up political agendas, and a growing interest in the science of functional foods is opening up many therapeutic possibilities.* (British Medical Journal, 2004)

The lobbying of industries, the sluggish rate at which medical insights are generally accepted and the emphasis on acute medicine instead of on preventive medicine are a few of the main reasons why we are still stuck with the outdated pyramid and plate.

In addition to the content, we can also say a few words about the shape of these food figures. I mentioned earlier that I find the plate a lot less clear than the pyramid. A pyramid is better, but why do most countries still use a pyramid or a plate? Could there be a better alternative?

The main problem with the current pyramid and plate is that they primarily say what foods you should eat, but they don't clearly state which foods you should avoid. The pyramid makes an attempt to solve this problem by adding a 'forbidden' section, a miniscule triangle at the top in which all the prohibited and unhealthy foods are put together. But there are so many more unhealthy foods than healthy foods! Sound nutritional advice should comprise putting unhealthy foods into clear view, so you know what you should avoid. Putting all those different kinds of unhealthy foods into one tiny top section, or even ignoring them altogether like in the food plate, ensures that people have no idea which foods to watch out for. In order to eat healthily, it is equally important to know which foods are unhealthy and which are healthy.

Another problem is that the pyramid and the plate offer hardly any alternatives. As an example, they say that you should eat less meat and more fruit and vegetables. That's pretty clear. But that is all there is. So what do you replace meat with? It is not explained anywhere that tofu or Quorn, for example, can be good meat substitutes, or that red meat can be replaced by poultry or fatty fish. There is definitely some pointing of fingers, but real solutions are hardly provided.

Yet another problem is the quantity. The pyramid and the

plate are divided into either groups or sections to provide you with an idea of how much of a certain food you are allowed to eat. For instance, the plate has a small section with fats, which is meant to tell you that you would do well not to eat too many 'unhealthy fats'. But there are quite a few food products which are actually rather healthy when consumed in small quantities, such as some food supplements and dark chocolate. In short, the tiny 'unhealthy' top of the pyramid and that small piece of the plate can just as well be filled with healthy foods you can eat in small quantities.

I started thinking about these problems and was already playing around with the idea of putting together a new kind of food pyramid that would be more consistent with new insights from biogerontology and the human metabolism. One day I was reading a book about nutrition in which Dr Roy Walford suggested his version of the food pyramid (almost every scientist or physician who writes a book seems to have his own version). This could be further improved. Suddenly I had an idea. I started drawing a picture that, to me, not only represented the healthiest nutritional method I could imagine, but also gave a clear picture as to what was healthy and what wasn't, and how the unhealthy diet could be substituted with other, more healthy alternatives. This figure became the 'food hourglass'.

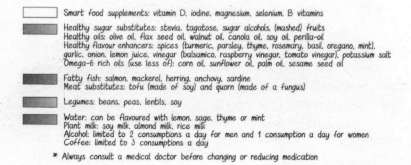

Smart food supplements: vitamin D, iodine, magnesium, selenium, B vitamins

Healthy sugar substitutes: stevia, tagatose, sugar alcohols, (mashed) fruits
Healthy oils: olive oil, flax seed oil, walnut oil, canola oil, soy oil, perilla-oil
Healthy flavour enhancers: spices (turmeric, parsley, thyme, rosemary, basil, oregano, mint), garlic, onion, lemon juice, vinegar (balsamico, raspberry vinegar, tomato vinegar), potassium salt
Omega-6 rich oils (use less of): corn oil, sunflower oil, palm oil, sesame seed oil

Fatty fish: salmon, mackerel, herring, anchovy, sardine
Meat substitutes: tofu (made of soy) and quorn (made of a fungus)

Legumes: beans, peas, lentils, soy

Water: can be flavoured with lemon, sage, thyme or mint
Plant milk: soy milk, almond milk, rice milk
Alcohol: limited to 2 consumptions a day for men and 1 consumption a day for women
Coffee: limited to 3 consumptions a day

* Always consult a medical doctor before changing or reducing medication

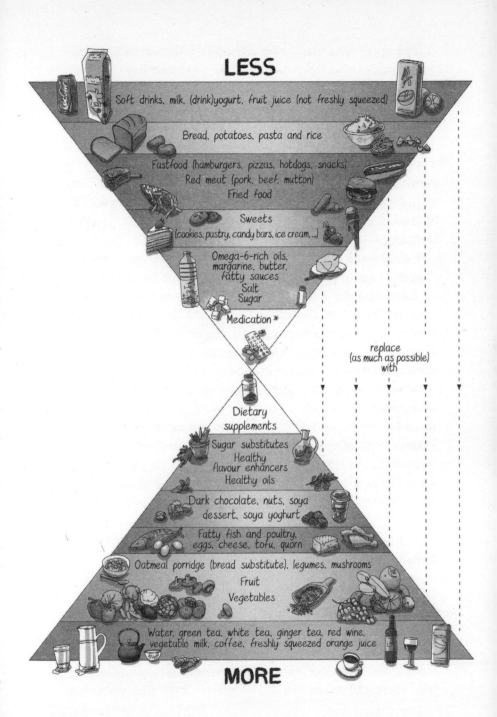

LESS

Soft drinks, milk, (drink)yogurt, fruit juice (not freshly squeezed)

Bread, potatoes, pasta and rice

Fastfood (hamburgers, pizzas, hotdogs, snacks)
Red meat (pork, beef, mutton)
Fried food

Sweets
(cookies, pastry, candy bars, ice cream, ...)

Omega-6-rich oils,
margarine, butter,
fatty sauces
Salt
Sugar

Medication *

replace
(as much as possible)
with

Dietary
supplements

Sugar substitutes
Healthy
flavour enhancers
Healthy oils

Dark chocolate, nuts, soya
dessert, soya yoghurt

Fatty fish and poultry,
eggs, cheese, tofu, quorn

Oatmeal porridge (bread substitute), legumes, mushrooms
Fruit
Vegetables

Water, green tea, white tea, ginger tea, red wine,
vegetable milk, coffee, freshly squeezed orange juice

MORE

Pyramids, plates and hourglasses **43**

The food hourglass automatically follows these seven basic rules:

- No – or as little as possible – bread, potatoes, pasta and rice.
- Substitute bread with oatmeal porridge (made with vegetable milk). Replace potatoes, pasta and rice with legumes (peas, beans, lentils, etc) or mushrooms.
- No milk or yoghurt. Replace these with vegetable milk (such as soya or almond milk) and soya yoghurt or soya dessert. Cheese and eggs are allowed.
- Replace red meat (pork, beef, veal, horse) as much as possible with fatty fish (salmon, mackerel, herring, anchovies and sardines), poultry (chicken, turkey), tofu or Quorn.
- Vegetables and fruit are the foundation of the food hourglass.
- A lot of water, a few glasses of green or white tea per day and at least one glass of freshly made fruit or vegetable juice. Coffee (maximum three cups a day) and alcohol are allowed (maximum two glasses a day for men, one glass for women).
- Taking 'smart' dietary supplements (iodine, magnesium, vitamin D, selenium, B vitamins).

An advantage of the food hourglass is that it clearly shows those foods which are unhealthy, so you have a better idea of what to look out for. Moreover, you can immediately see how to substitute the unhealthy foods with healthy alternatives. Another difference can be found in the 'top' of the lower triangle. Here you don't see those unhealthy foods that you need to avoid, but rather healthy foods that are at the top simply because you should eat smaller amounts of them, such as food supplements or herbs.

The food hourglass is based on solid scientific research from a diverse range of disciplines, such as endocrinology (the science of metabolism), biogerontology (the science of ageing) and preventive medicine. The food hourglass is also based on our evolutionary history. For example, our distant ancestors did not eat dairy products or bread for more than 99.9 percent of our entire prehistory; our brains were able to increase in size because we started consuming omega-3 fats; and our forefathers ate relatively little meat, contrary to what the paleolithic diet claims and unlike the storybooks that show our ancestors regularly hunting down a huge mammoth for the tribe's evening meal.

As stated earlier, the focus of the hourglass is not to lose weight, but more on slowing down the ageing process and reducing the risk of developing age-associated diseases. A genuinely healthy diet doesn't have weight loss as its goal, but does have the aim of reducing the risk of chronic (age-related) diseases. The weight loss comes automatically as a result.

The hourglass shape also symbolises time and how we can get a grip on time with our diet; how we can get our biological clock to tick a little slower. The sand that runs through the hourglass symbolises the passage of time and, likewise, the food that runs through our body gives an indication of how quickly we age. Those who eat healthily, live longer.

Some practical matters

The two triangles of the food hourglass are mirror images of one another. The upper, inverted triangle contains all food-stuffs we should try to eat as little as possible of; the lower triangle shows us the foods we should eat the most of.

At the same time, the food hourglass consists of two *symmetrical* triangles. This basically means that the sub-divisions or groups are each other's polar opposites. It is therefore quite clear to see for each group which unhealthy foods can be substituted by more healthy alternatives.

Let us begin with the base of the lower triangle, the first group. This group represents the healthiest drinks that we know, for example green tea, white tea, ginger tea, water or a glass of red wine. The base of the upper triangle, the highest group, is its polar opposite and in turn consists of those beverages that are less healthy, such as soft drinks and milk.

The same applies for the tops of the triangles, which are also each other's polar opposites. In the top of the lower 'healthy' triangle are dietary supplements. The top of the lower triangle does not contain 'forbidden' foods that are not recommended because they are located in the 'forbidden top'. This is the area with those nutrients that we take in the smallest of quantities. A few capsules of food supplements are smaller than the two litres of water or the various portions of vegetables that we should consume every day.

The colours inform us which foods are each other's polar opposites. You can see that the red group in the upper triangle, containing things such as red meat, matches the red group in the lower triangle. This means that we can substitute red meat with healthier alternatives in the red section of the lower triangle, namely fatty fish, white meat (poultry), eggs or tofu.

So how strict is the food hourglass as a 'diet' or dietary method? It is as strict as you want it to be. It is best to avoid the unhealthy food groups in the upper triangle as much as possible. On the other hand, the healthy foods in the lower triangle should be eaten as much as possible. People looking to lose an impressive amount of weight, or those who want to significantly slow down the ageing process, can decide to follow the guidelines of the food hourglass closely and stringently. This can be done by no longer eating bread, potatoes, red meat or dairy products. For most people, the food hourglass will be seen as a guide, which tells them at a glance what a healthy dietary pattern looks like. It is a dietary pattern that can be kept in the back of one's mind when preparing a meal, but needn't be followed to the letter. The food hourglass is here to serve as an informative guide, not to command and forbid.

SUMMARY

The **food pyramid** and the **food plate** are far from perfect because:
– they contain lots of non-recommended foods (such as red meat, dairy products and potatoes);
– they are too simple (not enough distinction between good and bad fats, red and white meats, white rice and wholegrain rice, regular tea and green tea or omega 6 and omega 3-rich oils);
– they are based on the classic American food pyramid, which has been strongly influenced by the dairy, meat and grain industries and which is a product of the US Department of Agriculture (and not the Department of Health).

The **food hourglass**:
– makes a clear distinction between good and bad fats, red and white meat, regular tea and green tea, omega 6 and omega 3-rich oils, etc;
– shows more clearly which foods are unhealthy (and doesn't cram all the unhealthy food in a small 'forbidden top');
– offers alternatives to unhealthy foods (through the symmetrical shape and colour composition of the hourglass);
– is based on insights not only into nutritional science, but also on those into medicine, biogerontology (the science of ageing), biochemistry and biology (our evolutionary history);
– focuses mainly on the prevention of chronic illnesses and age-associated diseases, while weight loss automatically follows.

So what, in a nutshell, is healthy food? According to my definition, healthy food is food that helps us obtain optimum weight and live longer, not only because we feel fitter and more healthy but also due to the activation of many different biochemical and metabolic mechanisms that cause us to age less rapidly. The following three chapters discuss the three major types of nutrients (sugars, proteins and fats) and explain what their impact is on the body, the metabolism and the ageing process.

You can download a coloured image of the food hourglass at www.foodhourglass.com

3

The three basic principles

Principle 1: Too much sugar is not unhealthy: it's very unhealthy

Everybody knows that too much sugar is unhealthy. However, sugar is even a lot more unhealthy than we think. Cavities in teeth or weight gain are actually the least of the problems to worry about when we eat too many sugars (carbohydrates). In fact, sugar causes:

1. accelerated ageing;
2. an increased risk of cancer.

Yet we cannot do without sugar. It is a little like oxygen: we need oxygen to stay alive, but oxygen also causes us to rust and age from within. It is the same with sugar: we get energy from sugar and we cannot live without it, but carbohydrates also accelerate the ageing process and increase the risk of cancer.

First, let's hear what professor Cynthia Kenyon from the University of California has to say. Professor Kenyon is one of the world's top researchers in the field of ageing. She performs genetic experiments on worms (*Caenorhabditis elegans*) and has unravelled a cellular mechanism that plays an important role in the ageing process: the metabolism of sugar and insulin. Because of this research, Professor Kenyon has become a renowned figure in molecular biology. Her research is published in leading scientific journals and she is a sought-after speaker at international conferences and congresses. In an interview, Kenyon states the following about sugars and nutrition:

'There's a lot of these [good]diets and what they all have in common is low carb – actually, low glycaemic index carbs,' she says. 'That's not eating the kind of carbohydrates where the sugar gets into your bloodstream very quickly [and stimulates the production of insulin].'

No desserts. No sweets. No potatoes. No rice. No bread. No pasta. 'When I say "no," I mean "no, or not much,"' she notes. 'I did it because we fed our worms glucose and it shortened their lifespan.'

'I have a fabulous blood profile. My triglyceride level is only 30, and anything below 200 is good.'

Kenyon is angered by the general lack of nutritional knowledge: 'It's a little bit embarrassing to say that scientists actually don't know what you should eat ... We can target particular oncogenes, but we don't know what you should eat. Crazy,' she says.

Does her dieting represent a return to scientists experimenting on themselves? 'I don't think so – you have to eat something, and you just have to make your best judgement. And that's my best judgement. Plus, I feel better. Plus, I'm thin – I weigh what I weighed when I was in college. I feel great – you feel like you're a kid again. It's amazing.' (*In Methuselah's Mould*, PLoS Biology, 2004)

It's clear: Professor Kenyon isn't very fond of sugar. Nor is she fond of potatoes, bread, pasta or rice. Because of her research, she has witnessed first-hand the significant role that sugar plays in the ageing process. But before proceeding any further, we should first discuss the precise nature of sugars in more detail and in particular their various forms, from sugar cubes to bread.

To understand what sugars or carbohydrates precisely are (and fats and proteins too) you have to know something about atoms, the basic building blocks of all matter around us. The brilliant physicist Richard Feynman considered the most miraculous scientific discovery that everything is composed of atoms. The entire world around us consists of these tiny particles, of which only 92 types exist in nature in a stable form (there are also roughly twenty rare types of atoms that occur very shortly in nuclear reactors, but these aren't included). The 92 types of building blocks become trees, drops of water, traffic lights, bodies and sugars. All 92 atom types are arranged according to 'weight', from the lightest atom, 'hydrogen' (H, element number 1) to the leaden and massive 'uranium' atom (U, element number 92). A block of iron consists of iron atoms, a block of gold of gold atoms and living organisms are mainly a combination of carbon (C), oxygen (O), nitrogen (N) and hydrogen atoms (H).

Atoms link together like building blocks to form molecules. A molecule is therefore each structure that is composed of two or more atoms. We can now proceed with sugars because naturally, these are also made up of atoms.

The scientific name for sugars is 'carbohydrates'. Sugars are composed of one or more *monosaccharides*. 'Mono' being the Greek name for 'one' and saccharide the Greek name for 'sugar'. Monosaccharides are the basic building blocks of all sugars. There are different types of monosaccharides. A monosaccharide molecule is usually hexagonal or pentagonal in shape. So a monosaccharide molecule consists of a hexagonal or pentagonal ring of carbon atoms (C) with a few oxygen atoms (O) and hydrogen atoms (H) that project outward from the angular ring of carbon atoms. The most common monosaccharides are glucose and fructose.

Glucose (a monosaccharide). C stands for a carbon atom; H for a hydrogen atom; O for an oxygen atom.

Fructose (a monosaccharide). C stands for a carbon atom; H for a hydrogen atom; O for an oxygen atom.

Dextrose (also known as grape sugar) is made up of glucose, thus from individual monosaccharides. Your blood sugar is also composed of glucose: when a doctor measures your blood sugar in reality it is the number of glucose molecules in a millilitre of blood that is being counted. Because fructose is commonly found in fruit, it's also known as *fruit sugar*.

These simple monosaccharides can bond to one another. When a fructose and glucose molecule bond together, they form a *disaccharide*. For example table sugar (or granulated sugar) is a disaccharide that consists of a glucose and a fructose molecule. The white table sugar that you spoon in your coffee or sprinkle over your pancake is made up of this type of disaccharide. The scientific name for table sugar is *sucrose*.

Two monosaccharides (glucose left and fructose right) bonded together form a disaccharide, namely sucrose, white table sugar.

Thousands of glucose molecules can link together like wagons forming a long train. When thousands of monosaccharides such as glucose bond together they produce long chains of glucose molecules: *starch*.

Thousands of glucose molecules bond together and form starch. (A side note: where the lines cross each other, e.g. at the junctions, there is a bond with a carbon atom (C), which isn't depicted in accordance with standard scientific notation).

Rice, potatoes, pasta and bread are all made of starch and thus consist mainly of long chains of glucose molecules.

Thus, grains of rice and strings of spaghetti consist primarily of sugar!

So far so good for our lesson in the atomic kitchen. It is also important to know one more thing and that is how the body absorbs these sugars. The gut cells in the intestinal wall can only absorb monosaccharides. They can only absorb free molecules of glucose or fructose, which they then release into the blood. That isn't a problem when you ingest a dextrose tablet because it consists primarily of free glucose molecules, which are absorbed directly into the blood through the gut cells. That's why dextrose is also given to sportsmen or people who feel faint and are in need of a quick boost of energy.

But this doesn't happen when you eat a sandwich because sandwiches contain starch. Starch can't simply or instantly be absorbed by the gut cells: starch is in fact made up of long chains of glucose molecules. Nature, however, has provided us with digestive enzymes to deal with this. Enzymes are proteins, tiny molecular machines that break down glucose chains until free glucose molecules remain, which can then be absorbed by the gut cells. Starch is broken down in the mouth, stomach and intestines into free glucose molecules that can be absorbed.

In this context, scientists refer to *fast* and *slow sugars*. Fast sugars are carbohydrates such as dextrose. Because they're made up of free glucose molecules, these sugars immediately enter the blood through the intestinal walls. Products such as bread or rice, consisting primarily of starch, contain slow sugars. The starch first has to be broken down into free glucose molecules in the digestive system. Because of this, it takes longer for glucose molecules, the elementary component of bread and rice, to be absorbed and that's why scientists refer to these as slow sugars. Ordinary table sugar formed by a glucose molecule and fructose molecule bonded together is a fairly fast sugar because it only requires a single 'snip' by a digestive enzyme to split the disaccharide into its individual elementary components of glucose and fructose, which can then be instantly absorbed into the blood through the intestines. So, if you take a large bite out of your biscuit (which contains a lot of sucrose), the glucose level in your blood will rise fairly quickly.

However, we're not there yet. We now know that various types of carbohydrates exist (monosaccharides such as glucose, disaccharides such as table sugar and polysaccharides such as starch) and we know that digestive enzymes are required to break down most sugars to release glucose molecules that can be absorbed into the blood. But what happens then? Roughly an hour and a half after a sugary meal, a high concentration of glucose molecules is present in the blood. These glucose molecules have to be absorbed by cells so that they can be 'burned' and converted into energy in order to keep the mechanisms of the cell functioning. Glucose molecules can only be absorbed by cells if another substance, a hormone, is present in the blood: insulin.

Insulin causes our cells to absorb glucose. Insulin is the substance that opens the gates of the cell so that glucose molecules can be absorbed. The pancreas secretes insulin when it detects a rise in sugar level in the blood (this already occurs during a meal). In sum: every time you ingest a sugary meal, the glucose level in the blood rises, prompting a subsequent increase in insulin levels, so the glucose can be absorbed by the cells.

The chain reaction continues however. What is important to know is that the high level of glucose and insulin in the blood also triggers the release of another substance, *insulin-like growth factor* (IGF). IGF is a type of 'growth hormone', an important substance that stimulates tissue growth. The 'real' growth hormone, which can be purchased from numerous online internet shops as a panacea against ageing, in fact causes the body to secrete more IGF so that all types of body tissue start to grow and develop. In short, IGF increases the production of proteins in the muscles as well as in many other tissues, resulting in the development of more compact and stronger muscles, a firmer skin and so on.

It's logical that glucose increases the IGF-level. In fact, a high level of glucose in the blood indicates that the person in question is indulging in sugary, energy-rich food. Which means, there is sufficient food available for growth. Hence, a sugar-rich meal causes an increase in the production of IGF so that the body can grow and become fully functioning.

Schematically:

Starch, table sugar or dextrose (grape sugar) → rise in glucose levels in the blood → insulin → *insulin-like growth factor* (IGF) → more growth and tissue development.

Why am I telling you all of this? Firstly, studies into ageing suggest that animals age faster when they ingest more glucose or when they produce a lot of insulin and IGF. Professor Kenyon could even genetically manipulate her worms in such a way that the cell's insulin and IGF switches functioned less efficiently. The worm's bodies were deceived into believing that there was less sugar, insulin and IGF present in the blood and because of this their life expectancy increased by a factor of three to six times when compared to normal.[1] Even the life expectancy of yeast cells, fruit flies and mice can be prolonged by reducing the sensitivity of insulin- and IGF-receptors.

This study also illustrates that one of the most acclaimed anti-ageing products ever, namely growth hormone, doesn't prevent ageing but in fact accelerates it. Growth hormone actually stimulates the production of more IGF, which accelerates ageing. Any doctor, who is a tiny bit aware of the research into ageing, knows that growth hormone is unhealthy when taken by a healthy person. However, Google 'growth hormone' and you immediately get hundreds of thousands of sites that praise this hormone as a panacea against ageing. This once again illustrates how distorted health advice can be. The negative role of growth hormone in the ageing process also explains why the life expectancy of smaller animals is on average longer than that of larger animals (within the same species of course). On average, smaller dogs live longer than larger dogs, as do smaller people than bigger people.[9] This is because, on average, small animals were less exposed to growth hormone (whereby they grew less and were therefore smaller).

Of course if someone ingests growth hormone then that person will often feel and look better. They will develop more muscle mass, skin will become firmer, fatigue will be reduced, libido can increase and so on. But, at the same time, the risk of accelerated ageing, cancer, diabetes, muscle pain and fluid retention increases.[10] Use of growth hormone makes you look better, but you age faster. Compare it to a car. Growth hormone is akin to replacing the engine of an old Beetle with that of a pristine, fully functioning engine from a Ferrari, which wears the car out far quicker and eventually destroys it.

Scientific studies reveal that the more or the faster someone grows, the quicker that person ages. That sounds logical. Growth hormone-like substances such as IGF ensure a more active metabolism whereby the body also deteriorates at a faster rate. But what perhaps is less obvious is the fact that sugar plays a significant role in that process. When someone eats sugary products, let's say an excellent Brussels' waffle, then the IGF-level increases, which in turn causes that person to age faster. That's why the more sugar that Professor Kenyon gave to her worms, the quicker they died.

Sugar, AGEs and diabetes

Insulin activation and IGF aren't the only means by which sugar can cause accelerated ageing though. Glucose also has a direct effect on our tissues which makes them age faster. Glucose molecules have the tendency to adhere to the protein in our bodies, causing these proteins to stick to one another. Proteins are the basic building blocks of our tissues. Muscle cells, stomach cells or epithelial cells: they all take their specific form and function from the proteins of which they're made.

Just like sugar is sticky in the kitchen on a macroscopic level, it's also sticky at the microscopic level. In the kitchen, sugar causes our fingers to stick together, in the body it causes proteins to stick to each other. Scientifically put, glucose forms *cross-links*, or bonds, between proteins. Cross-links are glucose-based compounds that are formed between two proteins that cause these independent proteins to stick to each other.

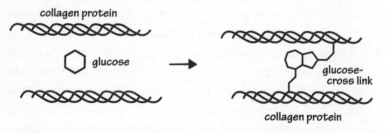

Glucose (sugar) adheres to proteins and forms bonds ('cross-links') between various proteins, so that tissues become more stiff or wrinkly. *Source: Johan Svantesson Sjöberg*

When these cross-linked proteins form in a particular tissue, the tissue becomes 'stiffer'. Cross-linking between collagen proteins in our skin for example, makes the skin less supple and less elastic, which creates wrinkles. That's why people who follow a low-sugar diet for a few weeks also have fewer-wrinkles in their skin.[11] When cross-linking occurs between the collagen and elastin proteins found in the walls of our blood vessels, the blood vessels harden. Our blood pressure rises as a direct result of this and this is a common symptom of old age. Hardened blood vessels have an increased risk of tearing or cracking. When this occurs in the brain, it's referred to as a brain haemorrhage or a stroke, yet another illness that elderly people have to contend with.

Cross-linking occurs everywhere in the body: for example, between the proteins in our kidneys reducing our renal function slowly as we get older. When cross-linking forms in the walls of our cells the immune system sometimes finds it difficult to recognise them as our own, aggressively attacking them as it would foreign bodies, resulting in an increased risk of auto-immune diseases such as rheumatism. Cross-linking occurs in cells from which our eye's lens is made; proteins clump together resulting in cataracts. It is no co-incidence that renal problems, rheumatism, cataracts, wrinkles, high blood pressure and strokes are all symptoms of old age.

The devastation that sugar can cause in the body is clear to see in those suffering from diabetes. In diabetes, the glucose level in the blood becomes too high, certainly after a meal. That's why diabetics have to contend with a range of illnesses associated with diabetes that are also common to old age such as an increased risk of cardiovascular disease, diminished sight, renal problems, erectile dysfunction, poly-neuropathy (nerve pain in the limbs), gastroparesis (delayed gastric emptying) and reduced bowel function because the nerves that control these organs become damaged by the sugar. In the West, diabetes is the leading cause of blindness, amputations, renal failure and kidney transplants. Amputa-tions are sometimes necessary because wounds tend not to heal well in diabetics, especially around the area of the feet (this is because sugar damages blood vessels, which results in poor blood circulation to the feet). These wounds can become easily infected owing to the fact that glycated white blood cells are no longer able to effectively combat bacteria, which has significant consequences.

The risk of cardiovascular disease is up to five times higher in people with diabetes. This is caused by sugar in the blood adhering to circulating cholesterol particles for example. As a result, cholesterol particles can more easily adhere to the smooth walls of blood vessels and form cholesterol plaques that eventually clog the entire blood vessel. If this occurs in a major artery in the heart, the result is a heart attack. When an artery in the brain becomes clogged, it leads to a stroke. Furthermore, sugar molecules adhere to coagulation proteins present in blood plasma. These coagulation proteins become even more sticky than normal, adhere at a faster rate to each other and accelerate the process known as the coagulation cascade, causing the blood to clot more easily and so to more rapidly clog blood vessels. In short, excess glucose in the bloodstream explains why 80 percent of people with diabetes die of cardiovascular diseases.

Diabetes isn't referred to by doctors as the accelerated ageing disease for nothing. This disease dramatically illustrates the influence of sugar on the body, from increased wrinkling to renal failure and heart attacks.

Sugar and cancer

Apart from accelerating the ageing process, sugar also plays an important role in the development of cancer.

Studies show that a person who indulges in a lot of sugary foods has a higher risk of developing cancer. How is this possible? Again, IGF plays an important role here. After ingesting a sugary meal, the glucose level in the blood rises and as a result, the body secretes insulin and IGF. As mentioned previously, IGF stimulates tissue growth and this is exactly what cancer likes to do: grow. Cancer occurs when certain cells grow rapidly and uncontrollably. What was first just one cancer cell, becomes millions and then billions, which together form growths or tumours. Substances such as IGF cause cells to grow faster, increasing the chance that they degenerate and become cancerous. Therefore IGF raises the risk of cancer and studies also show this. Women with high concentrations of IGF in the blood are seven times more likely to get cancer than women with lower levels of IGF.[12] Men who produce a lot of IGF are nine times more likely to have prostate cancer.[13–14]

Experiments with animals also show comparable results. A well-known study is the following. Mice were injected with a certain amount of cancer cells and subsequently divided into two groups by researchers. One group of mice were fed food containing a lot of sugar and the other group were fed foods containing very little sugar. After two months, 66 percent of mice that had ingested the large quantities of sugar died compared to 5 percent of the mice that had consumed considerably less sugar.[15] Toward the end of the study it was found that the mice that ingested less sugar were thirteen times less likely to die.

The relationship between IGF and cancer also explains why people with a certain type of dwarfism, called Laron dwarfism, almost never get cancer. They are small because they secrete less IGF which drastically reduces their chance of developing cancer (the risk is practically zero).[16] The reverse also applies: bigger people have an increased risk of cancer. Studies published in *The Lancet* indicated that for every extra ten centimetres (fractionally under four inches) of height (beginning at 1.52 metres, just under five feet) the risk of cancer increased by 16 percent.[17] And naturally, the very hyped anti-ageing growth hormone increases the risk of cancer, which can even be read in the patient information leaflet included in the medication packaging.

An increased IGF-level is however not the only reason why sugar promotes cancer growth. Cancer cells are highly dependent on sugar to grow. Cancer cells utilise a lot more sugar than regular cells. This is because of the *Warburg-effect*, which observes that cancer cells like to shut down their own *mitochondria*. Mitochondria are the power plants of the cell. Mitochondria burn sugars and fats, producing energy molecules such as ATP that stimulate all sorts of proteins in the cell, and thus practically maintain all intracellular processes. But mitochondria also produce *free radicals*, which are highly reactive molecules that damage various components of the cell. Mitochondria in cancer cells generate a lot more free radicals, because cancer cells are unable to maintain their mitochondria. In fact cancer cells expend all their time and energy on growth and division, and less on maintaining their mitochondria. Cancer cells are slovely cells and so are their mitochondria, which means they generate a lot of harmful free radicals. Cancer cells have found a solution to that problem: they simply deactivate their defective mitochondria, just like you throw away a faulty toaster because of the risk of fire. So where do cancer cells get their energy from? They are unable to burn sugars in the mitochondria so they do it outside of the mitochondria via another pathway: *glycolysis*. Glycolysis converts sugar molecules into energy. This process, however, is less efficient than burning sugar is inside the mitochondria. This means a lot less energy is generated per sugar molecule in the glycolysis-*pathway* compared to a sugar molecule that is burned in mitochondria. The result is that cancer cells have to utilise a lot more sugar to be able to survive. Thus, cancer cells are enormously fond of sugar.

PET scanners utilise this mechanism to detect cancers in the body. Before they are placed into the scanner, patients are injected with sugar molecules labelled with a light radioactive marker. The sugar molecules can be monitored via the scanner, which measures the radioactivity of the marked sugar molecules, their trajectory and where they accumulate. The tissues that absorb the most sugar are illuminated, indicating if the patient has bad luck, a cancerous growth.

Because sugar plays an important role in the onset as well as the maintenance of cancer, scientists are busy developing anti-cancer drugs that inhibit the metabolism of sugar. These drugs inhibit, for example, the aforementioned glycolysis pathway,[18] so that it is especially cancer cells that are affected.

Another reason why sugar enhances the risk of cancer is because sugar suppresses the immune system. Each day, thousands of cancer cells are formed in the body. The majority of these cancer cells are destroyed by the immune system, but if fate doesn't favour you, then a cancer cell can be missed and after many years develop into a cancerous growth. The immune system, especially the white blood cells that are the soldiers of the immune system, is the first line of defence in preventing cancer. However, sugar molecules adhere to the white blood cells, inhibiting their ability to perform their task effectively. The suppression of the immune system not only prevents the fight against cancer cells, but bacteria as well. The normal glucose level in the human body is between 75 and 110 milligrams per decilitre of blood. If the glucose level rises to a level above 120 mg/dl, then the ability of white blood to fight bacteria is halved.[15] Other studies indicate that people with a high blood sugar level (around 190 mg/dl) are twice as likely to die from blood poisoning as those with a lower blood sugar level (100 mg/dl).[19] Blood poisoning is caused by bacteria and the toxins that these bacteria release which is a dreaded hospital complication.

SUMMARY

Laboratory animals and people **age faster** and have a higher risk of **cancer** if they ingest a lot of **glucose**:

- glucose **activates IGF**, which increases the rate of growth and thus accelerates ageing and furthermore enhances the risk of cancer because IGF stimulates cell growth;
- glucose is the most important **fuel for cancer cells**;
- glucose creates **links between proteins** and causes symptoms of ageing such as:
  - wrinkles;
  - hardened blood vessels (and high blood pressure);
  - cataracts;
  - increased blood clotting;
  - nerve damage;
  - diminished renal function;
  - a weakened immune system.

These are all symptoms of **diabetes,** called an 'accelerated ageing disease', in which high glucose levels occur, especially after a meal.

The aforementioned findings also explain why:

- people with **a high level of IGF** in the blood are several times more likely to have cancer;
- laboratory animals that ingest **more sugar** die sooner from cancer;
- people with **dwarfism** almost never get cancer;
- the **bigger** someone is, the greater the risk he/she has of developing cancer;
- **growth hormone** causes accelerated ageing and a greater risk of developing cancer and diabetes;
- life expectancy of laboratory animals can be increased by modifying the **genes** that play a role in the insulin and IGF metabolism;
- **smaller** animals and people live longer than larger animals and people.

About bread, potatoes, pasta and rice

Sugar plays an important role in our health. Not so much because sugar causes cavities in our teeth but because it accelerates the ageing process and stimulates cancer growth. It is important to limit the amount of sugar in our diet, but not too much as is the case with the classic Atkins diet and many other very low-carb diets. In these diets, the use of carbohydrates is discouraged, yet ironically this causes more toxic sugar products to be produced. We need sugar just as we need oxygen. If we indulge in sugar products it is important to indulge in those products that release sugars slowly into the bloodstream so that high peaks of sugar, insulin and IGF are avoided. Products that yield low sugar peaks are products with a *low-glycaemic index* (as we shall soon see). Wholemeal bread has a lower glycaemic index than white bread because it produces less pronounced sugar peaks in the blood.

I will refer back to the interview with Professor Kenyon. Professor Kenyon says that she doesn't eat desserts or confectioneries any more. That sounds logical because they are bursting with sugar. However, she has also stopped eating bread, potatoes, rice and pasta as well. These foods also contain sugar in the form of starch. They are slow sugars, which mean they are healthier than the fast sugars found in a waffle or a cake, but they are, and remain, sugars.

Now I have come to an important point: if you want to lose weight in a spectacular way or live very healthily, then you must eat no or very little bread, potatoes, rice and pasta. You can do this for a limited period, to lose weight, or even stick to this during your entire life, as Cynthia Kenyon does. To obtain a great result you could stop eating bread, potatoes, rice or pasta for a few weeks, and after that period you could just eat small quantities of these foodstuffs. The results are impressive and it is even possible to 'cure' type-2 diabetes. That is to say, the cure lies in the fact that diabetes patients will not need to use as much insulin as previously, or no insulin at all, (of course, this needs to be done under a doctor's supervision).[20-23]

Type-2 diabetes (also known as adult-onset diabetes) is the most prevalent form of diabetes. This disease is caused primarily by excess weight and old age. Being overweight especially prevents the cells in the body from being receptive to insulin. They are *insulin-resistant*, as doctors say. Because of this the sugar in the blood isn't absorbed as efficiently by the liver, muscle and fat cells that would normally absorb and store large amounts of sugar. As a result the sugar continues to circulate in the bloodstream, damaging nerve cells, eye cells, kidney cells, blood vessel walls and so on. Type-1 diabetes on the other hand is a very different disease. Type-1 diabetes is much rarer than type-2 and presents more commonly in young and thin people. This disease arises because the immune system attacks the body's own beta cells in the pancreas that produce insulin. Because of this, people with type-1 diabetes cannot produce their own insulin and when they eat a meal, the sugar remains in the bloodstream because there's no insulin to open the gates of the cells so that the sugars can be stored.

It's type-2 diabetes that plagues the West and many emerging developing countries. This is mainly due to bad eating habits such as eating too many sugars. Some people claim eating sugar doesn't cause diabetes and this is obviously both true and false. A thin person who eats sugar will not develop diabetes. But if he continues to overindulge in sugar and as a result of that, becomes fat, he may develop type-2 diabetes because being overweight is the most important risk factor. So in the long term, sugars are dangerous with regard to developing type-2 diabetes.

Many people believe that 'fattening' sugars are found chiefly in biscuits, tarts, soft drinks and cake, but they miss the most important daily source of these sugars, a source that most people indulge in large quantities two or three times a day: bread, potatoes, pasta and rice. By avoiding these foods, or eating a lot less of these foods, you not only lose an impressive amount of weight, you slow down the ageing process and can even reverse type-2 diabetes.

As a doctor, I was taught that type-2 diabetes was a chronic disease (once diabetic, always diabetic) and that once people began injecting insulin, it was a lifelong necessity. I was hugely surprised when I read the first reports documenting diabetes patients who had lost considerable amounts of weight after cutting out bread, potatoes, pasta and rice. After a few weeks they hadn't just dramatically lost weight, they didn't have to inject insulin any more.

Of all the diets that these diabetes patients had tried, this was the only diet that had such a far-reaching effect. I've seen similar results with my own diabetes patients. The first was an uncle of mine, when I was still a medical student. I advised him to avoid bread, potatoes, pasta and rice. For many years he had injected himself three times a day with insulin and, despite several frantic diet attempts, his weight kept increasing along with his blood pressure and eye problems. Of course this meant he had to inject more insulin. But, by cutting out bread, pasta, rice and potatoes, he not only lost an incredible amount of weight, after a few weeks he didn't have to use insulin anymore: his blood sugar levels remained normal.

That patients are able to drastically reduce their insulin use with this diet is particularly interesting. Although, diabetics can't do without it, insulin is in fact a double-edged sword for these patients. Insulin, like IGF, works like a growth hormone; it causes body tissues to grow, especially fatty tissue. Insulin causes people to get fat. Being fat also happens to be the largest risk factor for type-2 diabetes. In short, every diabetes patient who injects insulin is caught in a vicious circle: he becomes fatter from using insulin and advances his diabetes, for which he needs to inject more insulin and thus become even fatter and so on. Studies also show this: injecting insulin turns back the clock for a couple of years, but the diabetes still progresses. This is graphically illustrated in the form of two parallel lines: diabetes increases equally quickly in patients who either do or don't inject insulin; the line representing insulin-injecting patients is only positioned lower on the graph because those patients only briefly turn back the clock. In the meantime the clock continues to tick just as fast.[24]

Cutting out (or reducing) bread, potatoes, rice and pasta is a dietary pattern that is not generally known yet but during my internship as a medical student I came across patients who had already discovered a similar diet. They were in the minority but they were particularly satisfied with their weight loss and improved health. In view of the damage sugar can cause to the body, that's not so surprising.

This diet is not widely known yet, not even in medicine. The standard diet on the diabetology ward in many hospitals is still a low fat diet (very ineffective on the long term) or even worse a high-protein one (a bit like the Atkins diet). In the short term, this protein-rich diet often causes enormous loss of weight, which gives the desired result during a short hospital stay. However, the downside is that it is an unhealthy diet[25], especially for diabetics, and one that patients find difficult to sustain over a long period of time. Instead of following a high-protein diet, patients should be taught to replace bread and potatoes with other foods. Moreover, this diet, unlike the protein diet, can be followed for a lifetime and can ensure that many patients no longer need to inject insulin. It is a diet that scientific research shows may even extend the lifespan of healthy organisms.

Some will wonder whether foregoing bread, potatoes, pasta and rice is a form of *ketogenic* diet: a diet that causes the body to produce *ketones*. Ketones are substances the body generates when it receives too little sugar through food. The brain, which requires a lot of sugar in order to function, will then start to use the ketones as its primary source of food. The Atkins diet is also a type of ketogenic diet because carbohydrates are practically banned, especially at the start.

But the dietary pattern described in the food hourglass isn't a ketogenic diet. Products such as bread, potatoes, pasta or rice are indeed discouraged. But fruit and foods that contain starch such as beans, peas or other kinds of vegetable are not. These sources supply you with numerous sugars in a much healthier way, as we shall see.

I can hear you thinking, 'No – or very little – bread, potatoes, pasta or rice? What *can* I eat then?' Starchy products form the basis of almost all our dishes. No more bread? No more potatoes to accompany my hot meal? Impossible! Yet, in many non-western cultures a lot fewer products that are high in starch are consumed. Moreover, bread, potatoes, pasta and rice are only recent discoveries. They have only been consumed for a couple of thousand years. The human species is 180,000 years old. So for more than 170,000 years people had to do without bread, potatoes, rice or pasta. The potato, nowadays the staple diet of practically every hot meal, was only introduced to Europe halfway through the sixteenth century. In other words, evolution built our bodies to survive without these 'contemporary' starch products. However, this still doesn't answer the question of what you can eat if these products are eliminated. Later in this book we will tackle this question extensively. For now, suffice to say that for breakfast, bread should be replaced by oatmeal porridge and, for example, a bowl of strawberries, blueberries, a pear, dried fruit, walnuts or a piece of dark chocolate. Such a meal is delicious and particularly healthy. I propose one exception to the 'no starch rule' and that is oatmeal porridge, which is made from oatmeal, a grain, and milk. It is important to use vegetable milk, like soya milk. Oatmeal porridge can serve as a replacement for bread in the morning. In the evening, potatoes, pasta or rice can be substituted with an extra portion of boiled or steamed vegetables, or by peas, beans or mushrooms. There are many delicious recipes that don't involve potatoes or pasta. Prehistoric man could do it, early sixteenth-century Europeans could do it, and Professor Cynthia Kenyon can do it. You can, too.

This book also addresses how to replace sugar with healthier alternatives. Those who have already thoroughly studied the food hourglass know that sugar can be substituted with stevia or sugar alcohols. In short, there are plenty of alternatives to replace the excess sugar that we consume nowadays.

SUMMARY

There are two ways of getting started with the food hourglass.

Strict interpretation: no (or very little) bread, potatoes, rice, and pasta, for example, for eight weeks, or for an unlimited period of time.

Suitable for:
- diabetes patients (under medical supervision);
- people who have trouble losing weight.

Example: John has type-2 diabetes. Once a week, he eats a meal with brown rice, but for the rest he doesn't eat bread, potatoes or pasta. After eight weeks he decides whether or not to switch to the 'less strict interpretation'.

Less strict interpretation: a lot less bread, potatoes, rice or pasta (for a certain period, or throughout your whole life).

Suitable for:
- people who want to lose weight;

- people who want to live a healthier life, or want to reduce their chance of getting ageing diseases.

Example: Susan eats two meals a day that don't contain bread, potatoes, rice or pasta.

So much for the carbohydrates. Now we must turn our attention to proteins and fats.

Principle 2: Watch out for proteins (and high-protein diets)

Proteins are the third source of calories after sugars and fats. Proteins are fascinating structures; they perform practically all the processes in cells and form the building materials of our tissues; they are the workhorses as well as the building blocks of the cell. Proteins are actually tiny clumps of atoms comprised of a few hundred to many tens of thousands of atoms, mostly ten nanometres in diameter (a nanometre is one millionth of a millimetre).

All proteins are made up of amino acids. Like sugars, amino acids link together to form long chains. An amino

The basic structure of an amino acid, the building block of all the proteins in our body. H stands for a hydrogen atom; C for a carbon atom; O for an oxygen atom; N for a nitrogen atom.

acid chain is called a protein. A long chain folds on itself in a particular way to create a specific form.

The basic structure of each amino acid consists of nine atoms that are always grouped in the same arrangement (see illustration).

The R-group consists of a particular group of atoms. This group determines the type of amino acid. There are 20 different amino acids in the body, and therefore 20 types of R-groups. An R-group can be made up of 1 atom, such as hydrogen (H). In this case, the amino acid is known as glycine. An R-group can also contain about twenty atoms, such as in *tryptophan*, the largest amino acid. Listed below are the twenty different amino acids present in the body, which build up all our proteins.

The order in which amino acids are linked together determines the type of protein. One protein can consist of glycine-arginine-tryptophan, whereas another can start with tryptophan-tryptophan-arginine- ... and so on. You can naturally achieve a practically infinite number of combinations. With three amino acids you can create 27 different types of protein chains, with ten amino acids you already have ten billion possible combinations.

The 20 types of amino acids listed above make up the approximately 100,000 different proteins in our body. Some proteins consist of a few dozen amino acids, such as insulin that's composed of a chain of 51 amino acids. Other proteins are giant proteins, such as *titin*, which consists of 30,000 linked amino acids and is an important protein in muscle tissue.

The order and type of amino acids determine the form of the protein. The atoms in the amino acids are positively or negatively charged; because of this, certain amino acids will attract or repel each other. The long chain of amino acids will start to fold onto itself until it achieves a certain form.

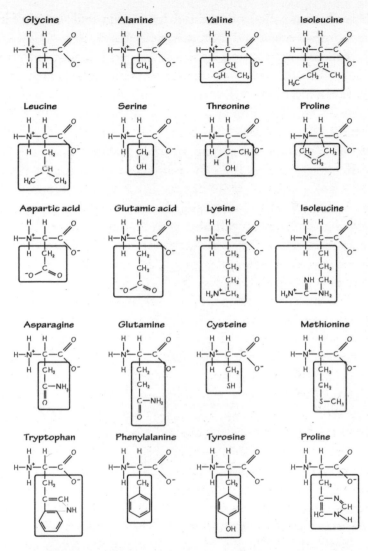

The 20 different types of amino acids present in the body and which compose our proteins. The R-group, which determines the type of amino acid, is framed.

Key: hydrogen atom (H), carbon atom (C), oxygen atom (O), nitrogen atom (N), sulphur atom (S).

Note: a carbon atom can be found at each angle of the hexagons (this atom isn't shown in accordance with standard scientific notation).

This characteristic leads to each type of protein having its own distinct form, and that form determines the function of the protein.

At the molecular level some proteins look like a hollow cylinder. They function as a 'gate' or 'door' in the cell walls:

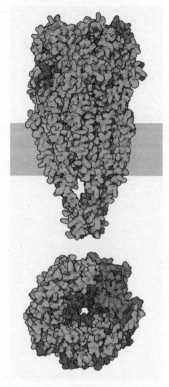

This channel protein allows certain charged atoms, such as sodium, to pass into the cell from outside. Each sphere represents an atom. *Source: David S. Goodsell, rcsb Protein Data Bank*

There are also proteins that resemble a hive; they function as a repository for metals such as iron:

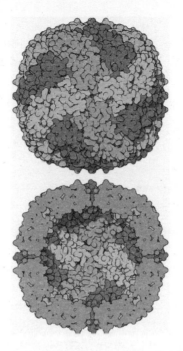

A ferritin protein is kind of like a round ball, where iron atoms are stored. Each little sphere is an atom. Ferritin is a type of hive where iron is stored. *Source: David S. Goodsell, rcsb Protein Data Bank*

Other proteins are long chains that can contract; this makes it possible for our muscles to move our limbs. Other proteins, such as collagen, focus on building our tissues. Collagen are long protein chains found between cells, such as our skin cells. Approximately a quarter of all proteins in the body are collagen proteins.

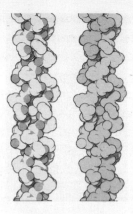

Two collagen proteins. Each little sphere represents an atom.
Source: David S. Goodsell, rcsb Protein Data Bank

One final thing that is important to know is how the body digests and absorbs proteins. This occurs in the same way as sugars, namely by breaking the long chains into smaller pieces. When we eat proteins, they are broken down by digestive enzymes until only individual amino acids remain. The amino acids are absorbed by the gut cells, released into the bloodstream and transported to the cells in our body. The cells absorb the amino acids and link them back together to form essential proteins. So the gut can only absorb free amino acids.

Proteins are the building block and workhorses of the body. All our cells contain a 'skeleton' that is made up of long chains of proteins. The cells in our muscles are full of long parallel chains of proteins that contract, which allows our muscles to contract so that we can laugh, dance or applaud. Almost every function in the body is performed by proteins: they are the gates in the walls of our cells, the repositories of the cell, the enzymes that break down substances or create new ones.

Meat (muscles) consists of lots of protein. That's why bodybuilders drink protein shakes or amino acid shakes to absorb a lot of raw materials (amino acids), produce proteins in the body and develop stronger and bigger muscles.

Eating a lot of high-protein meals may seem like a healthy option, especially if you want to look strong or muscular. But that isn't the case: too much protein is not healthy and the more proteins or amino acids you eat, the faster you age.

Fruit flies given a lot of amino acids live shorter lives,[26] especially if they are given essential amino acids that the body cannot make itself. Mice, put on a low-protein diet, age slower and live longer. But why is that?

Proteins encourage growth but as we've already seen, growth means ageing. In fact, all protein that is made becomes damaged over the course of time. Sugar molecules adhere to the proteins or free radicals damage or 'oxidise' the proteins, altering their structure so they become dysfunctional. These old, damaged proteins have to be broken down by the cell. But that is not always successful because worn proteins tend to adhere to one another and become unable to be broken down. This accumulation of proteins in the body becomes our final downfall. Many diseases associated with ageing are, in essence, diseases where protein accumulates either within or outside cells, suffocating the cells and resulting in their death. The standard case in point is Alzheimer's disease. With this disease, proteins accumulate and clump together within and outside brain cells so that the brain cells get overwhelmed by indigestable protein clumps in such a way that they die off. Another disease associated with ageing is *amyloidosis*. This is a disease to which even those over the age of 100 succumb, even after surviving cancer and cardiovascular disease. Amyloidosis occurs because proteins precipitate in and around organs and blood vessels until the organs and blood vessels become brittle and weak.

Therefore, too much protein is unhealthy. An experiment illustrating just how unhealthy was included in a study published in *Nature* in which groups of rats were given different diets with varying amounts of protein. One group of rats received a diet with 10 percent protein, another group had food that contained 22 percent protein and a third group was fed a diet with 51 percent protein. The rats that ingested the diet with 51 percent protein were four times more likely to develop heart disease and twice as likely to suffer kidney disease and prostate problems.[27] Kidneys are especially vulnerable to proteins. Proteins broken down in the body form amino acids and these are further broken down into nitrogen compounds excreted by the kidneys. This places a considerable strain on the kidneys and is why kidney patients are often put on low-protein diets.

Disease	Percentage protein			Free choice of in the dietprotein-intake
	10	22	51	
Heart disease	11	42	48	67
Kidney disease	38	56	73	90
Cancer	26	29	28	62
Disease of the prostate	5	10	12	62
Total	80	137	161	281

Source: Dietary preference and diseases of age. Nature, 1974

What is interesting about this study is the inclusion of a fourth group of rats. They were not put on a diet with a strict percentage of proteins. They were allowed to 'choose' how much protein they ingested. The risk of heart disease, cancer and kidney disease shot up among this group. These rats had double the occurrence of cancer, almost three times the occurrence of kidney disease and six times more heart disease when compared to the rats of which the diet consisted of only 10 percent protein.

*   *   *

People who follow popular high-protein diets like the Atkins diet 'know what to expect', to paraphrase Dr Roy Walford, an expert on nutrition and ageing.[28] According to these diets you have to eat a lot of high-protein meals as a substitute for carbohydrates (sugars). People on this diet certainly lose weight but they also lose good health.

Something similar happens with the popular classic paleo diet as well. According to this diet we should eat more like our caveman ancestors. This diet works on the basis that you should eat a lot of meat and is characterised by dishes with a huge piece of meat in the centre with some vegetables on the side.

I completely support the reasoning behind the paleo diet, namely that we should eat as our ancestors did a hundred thousand years ago. Our contemporary bodies are the same as they were back then and our digestion and metabolism have been designed by nature to digest food in practically the same way as they did a hundred thousand years ago. The problem with the paleo diet however is that proteins in the form of meat form the basis of the paleo-food pyramid. Meat as the basis of a food pyramid! Apart from the fact that this is particularly unhealthy, our ancestors probably didn't eat a great deal of meat so actually there's not much paleo in the paleo diet at all. It was not easy to regularly fell a mammoth, catch a quick gazelle or even skewer a skittish rabbit. The real paleo diet consisted mainly of vegetables, fruit and nuts, with a little meat: a diet upon which the food hourglass is based.

We have already observed that when rats, mice and fruit flies ate a lot of high-protein foods their life expectancy decreased and they were increasingly susceptible to numerous diseases. Other studies also show that too much protein isn't healthy for people.[29] Proteins place a considerable strain on the liver and kidneys and for this reason people with renal and liver failure are put on low-protein diets. Animal proteins acidify the blood and increase the risk of osteo-porosis (brittle bones).[30]

People who eat a lot of meat also have an increased risk of getting cancer and heart problems. A study in which 91,000 women took part showed that women who eat meat every day were twice as likely to get breast cancer as women who eat meat less than three times per week.[31] According to a study published in the New England Journal of Medicine it appears that people who eat beef, pork or mutton/lamb every day had two and a half times more chance of getting bowel cancer than people who eat meat less than once a month.[32] Scientists at Harvard University have conducted a study among 120,000 participants and have found a clear connection between eating red meat, getting cancer, and risking a heart attack.[33] With each daily portion of red meat a person eats, the chance of getting a heart attack increases by 20 percent. According to another study, among 450,000 Europeans, people who consumed a lot of meat (for example, two sausages and a slice of bacon per day), had a 44 percent bigger chance of dying during the course of the study than people who ate 20 grams of meat or less.[34]

In addition, proteins aren't always completely broken down in the intestines, certainly if you ingest a large amount through meat. These undigested fragments of protein (peptides) float around in the niches of the intestines and, because they were indiscriminately broken down by our digestive enzymes, the intestinal immune system recognises them as foreign objects. Because of this, the immune system can become overactive and we run the risk of developing autoimmune diseases such as coeliac disease (gluten sensi-tivity), asthma, eczema, inflammatory bowel disease and so on.[35]

And then there is *rapamycin*, a substance that can actually prolong life expectancy by inhibiting protein production in the body. Rapamycin caused quite a stir back in the summer of 2009 because it actually prolonged the life expectancy of mice. The study was published in *Nature*, not exactly an obscure medical journal, which shows that an antioxidant can extend the life of a dubiously sick mouse. According to this *Nature* study, the life expectancy of (already old!) mice could be considerably prolonged by approximately 38 percent for female mice and 28 percent for male mice.[36] Rapamycin is therefore one of the few substances that can truly prolong life. How does rapamycin work? Rapamycin inhibits mTOR (mammalian Target of Rapamycin), an important protein present in all of our cells that stimulates the production of proteins. mTOR activates the production of proteins in the cell and the more mTOR that is activated, the more proteins that are made in our cells, the faster we grow and the more muscular we become ... and the faster we age. Rapamycin inhibits the production of protein and ensured that the mice became a whole lot older. It's also interesting to note that mTOR is not only activated by amino acids, but also by sugar and insulin in the blood. In short, sugars and amino acids (and proteins) are substances that stimulate growth, but shorten life expectancy.

An overabundance of scientific research shows that too much protein is unhealthy. Animal products that are rich in proteins (such as meat) increase the risk of getting cancer and heart disease, and even diabetes.[37] It is also remarkable that interviews with centenarians demonstrate that these hundred-year-olds often have eaten very little meat.

Of course, meat can be unhealthy due to other causes than just the proteins. This is because meat, and especially red processed meat, such as bacon, sausages or salami, contains preservatives, unhealthy fats, colorants, and lots of salt.

Even so, proteins appear to play a part too. There is a direct link between proteins and the chance of contracting ageing diseases, such as diabetes. Per 5 percent increase in the ingestion of proteins (out of the total amount of calories) the risk of getting diabetes increases by 20 percent.[25] And if a lot of amino acids circulate in the blood, our metabolism will function less well, and our cells will become increasingly insulin-resistant (they will process sugar less well).[39–40]

Despite these insights, protein-rich diets are very popular. There are several reasons for this. Thanks to protein-rich diets you can lose a lot of weight in a short period of time, which leads people to think that they are doing a healthy thing. But the weight loss comes about in an unhealthy way. Besides that, in the long term it isn't very efficient: after six years, 90 percent of the people who follow a protein diet weigh at least as much as when they started.

Another reason why many people stick to proteins is that some studies do indeed demonstrate a beneficial effect to people's health. An example: if you give people amino acids, they will have a little more stamina or a better muscle metabolism. Which makes sense: in the short term, proteins can make people a bit stronger or fitter (that is why bodybuilders use them), but in the long term, too much proteins are unhealthy, because the amino acids actually activate all sorts of ageing mechanisms, such as the mTOR mechanism.[41–42]

Yet another reason (and probably the most important one) for the popularity of protein diets is the fact that they bring in a lot of money. Indeed, people need to ingest such large amounts of protein that they have to buy extra proteins in the form of protein drinks, protein powder or protein bars. This can cost tons of money, and will produce a lot of cash for the people who invent or tout such diets.

Does this mean that we should no longer eat meat and become vegetarians? Not really. Vegetarians do live longer than the average westerner, but not all vegetarians live to be 120 years old. Meat contains numerous substances that we need such as zinc, vitamin B12, carnosine, iron and so on. Therefore, we can eat meat, but in lower quantities than our society is used to. For example, don't eat more than the size of the circle you form when you place the tip of your thumb and index finger together a day. Or reduce the amount of times you eat meat this quantity to every other day.

Of course you shouldn't eat too few proteins either, because that will weaken your immune system and make you more susceptible to infections. People with a serious protein deficiency develop weak muscles and heart problems. It is therefore important to consume sufficient proteins, but preferably in smaller quantities than is generally eaten. I'll come back to this when we discuss the food hourglass in detail.

SUMMARY

Too many proteins accelerate ageing:
- proteins cause increased **growth**, which results in faster ageing;
- proteins continuously **clump together** within cells, as well as outside them, causing them to become 'choked-up' and damaged. This process plays a part in:
    - Alzheimer's disease;
    - Parkinson's disease;
    - amyloidosis (a disease in which proteins precipitate everywhere in the body);
    - general symptoms of ageing (such as diminished sight, hearing, deteriorating heart, kidney, and liver function because the cells are clogged up by protein clumps);
- medication such as **rapamycin** increases the lifespan of laboratory animals because this substance inhibits protein production;
- mice, fruit flies and other laboratory animals that eat **less proteins** or amino acids live longer;
- people who reach **very old age** often eat very little high-protein food such as meat.

Continued overleaf

An **increased intake of proteins or meat** also causes:
- an increased risk of **osteoporosis**;
- a strain on the **liver and the kidneys**;
- an increased risk of non-colonic **cancers** such as breast cancer;
- **poor digestion** in the intestines, whereby the 'rotting' of protein-rich products generates toxic substances in the intestines (increased risk of colon cancer) and fragments of undigested protein can activate the immune system (increased risk of auto-immune diseases such as gluten sensitivity).

Despite the fact that it's a very popular form of diet, **protein-rich diets** are unhealthy even though they promote weight loss.

However, a (serious) **protein deficiency** can impair health. Meat also contains **essential substances** such as zinc, iron and carnosine, which are important to health.

Principle 3: Fat is healthier than you think

It is a paradox: for forty years Americans have been eating less fats, but they are still becoming fatter. Through numerous health campaigns, the American people were informed that the consumption of fats is unhealthy: it makes you fat, clogs blood vessels and causes heart attacks. The campaigns were effective: Americans are eating less fat and this trend is still continuing. But since 1970 the number of overweight people has doubled, the number of diabetics has tripled, and the number one cause of death is still cardiovascular diseases. So let me tell you something that may contradict every piece of health advice that you have ever received about fats: most fats don't play an important role in cardiovascular disease. Yes, you read that right. A meta-analysis of 350,000 people confirmed the suspicions of many researchers that even the infamous *saturated fats* don't play an important role in cardiovascular disease (further on we explain the nature of saturated fats).[43] But don't saturated fats raise cholesterol level? Certainly, but there's good cholesterol as well as bad cholesterol. Fats raise the level of both, so there isn't an imbalance.

In addition, the role of fats in being overweight seems to be less important than that of sugar. At first, this can seem quite strange because obese people do hold more fat and because obese people have more fat, then fats must be the evil-doer. But it's not that simple.

In a study published in *The New England Journal of Medicine,* more than three hundred test subjects were put on three different diets: a low-fat diet, a Mediterranean diet (with lots of vegetables, little red meat, but of course, the obligatory pasta products) and a low-sugar diet. The findings were as follows: the people with the best cholesterol ratios in their blood were not the ones on the low-fat diet, but those who had followed the low-sugar diet. That's interesting because cholesterol is always associated with fats while the connection between sugar and cholesterol is less known. In addition, the people on the low-sugar diet lost twice as much weight as those on the low-fat diet.[44] What a blow to all those diets that say you have to reduce your fat intake in order to lose weight (because 'fats contain so many calories'). What this study also shows is that the famous and highly commended Mediterranean diet can be much healthier. After all, it contains a lot of carbohydrates in the form of bread and pasta.

It is especially the diets that contain few *fast* sugars that appear to be more effective and healthier than low-fat diets. This is because it's the fast sugars that cause high sugar spikes in the blood, which isn't healthy. As we shall see later on, the glycaemic index is the measurement for gauging the height of these sugar peaks. That is why diets that contain few fast sugars are called 'low glycaemic index diets'.

A large study, conducted by the famous Cochrane institute,

found that low glycaemic index diets are healthier for the cardiovascular system than low-fat diets, and they also make you lose more weight.[45] The latter aspect is all the more remarkable in that the low-fat diets also restricted the amount of calories (so you were not allowed to eat too much), while the low glycaemic index diet let you eat as much as you wanted, and you still lost more weight!

Another interesting study let groups of patients adhere to various diets: one group had to stick to a low-fat diet, while another group got a low glycaemic index diet. The results show better cardiovascular parameters (better cholesterol ratio, fewer fats in the blood, and a better insulin sensitivity) for the people who adhered to the low glycaemic index diet, compared to those who followed the low-fat diet.[46] This study also disproved the classic, widespread notion that a calorie always remains a calorie, whatever its source. Indeed, in this study, all the test subjects consumed the same amount of calories, but the subjects who followed a low glycaemic index diet had healthier blood vessels than those who followed a low-fat diet. The conclusion is that sugar plays a more important role in cardiovascular disease than most fats. Not only are diabetics five times more likely to be at risk from cardiovascular disease, but even in healthy test subjects there is a strong link between sugar intake and heart attacks. According to research in which 15,000 women took part, it was found that women who ate meals with a high glycaemic index (sugar products that cause large sugar and insulin peaks in the blood) were 80 percent more likely to be at risk from cardiovascular disease than women who ate meals with fewer fast sugars.[47] Or, in the words of Professor David Ludwig, director of the obesity programme in the Boston Children's Hospital, 'The next time you eat a piece of buttered toast, consider that butter is actually the more healthful component.'

What a lot of research tells us is that sugars play a particularly important role in cardiovascular diseases. Nevertheless, it is still advantageous to discriminate between healthy and unhealthy fats. *Trans fats* are extremely unhealthy fats that do indeed increase the risk of a heart attack. These fats are commonly found in fried foods and industrially-produced confectioneries such as biscuits, waffles or cakes. The other so-called 'unhealthy' fats, the renowned *saturated fats*, do not play such an important role in cardiovascular diseases; contrary to what was previously thought (in fact, sugar seems to be the much more important factor). There are also healthy fats, such as *omega-3 fatty acids*, that can actually offer extra protection against cardiovascular disease. In addition, these fats not only reduce the risk of a heart attack, but also reduce the risk of inflammatory diseases and depression. To understand the nature of these healthy fats, we have to examine fats in more detail. Each kitchen is in fact a small laboratory in which the molecular structure of food plays a crucial role in our health. The next section on the molecular structure of fats is the final, more difficult part of the book.

Lipidology

What is a fat? A fat is a molecule. Trillions and trillions of these fat molecules come together to form butter or oil. But the basic unit of these substances, the fat, is a molecular structure.

A fat molecule is made up of a few dozen atoms and always has a well-defined structure, namely a 'head' with multiple (two or three) 'arms', the *fatty acids*. The arm or fatty acid consists of a long chain of carbon atoms (C) on which hydrogen atoms (H) adhere. The head consists of a group of atoms that are known collectively as *glycerol*. You can actually picture a fat as a disabled squid with two or three arms, depending on the type of fat. The head is the glycerol group and the arms the fatty acids.

When we eat fats, they have to be broken down by

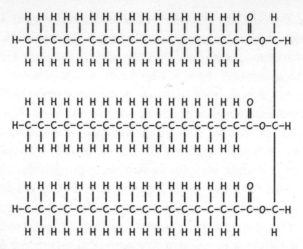

The molecular structure of a fat. The three arms (fatty acids) of the fat are on the left. The head of the fat molecule is on the right and consists of 3 carbon atoms (C) and a few oxygen (O) and hydrogen atoms (H). Lots of fats in our bodies have only two instead of three arms. *Source: Kimball's Biology Pages (http://biology-pages.info)*

digestive enzymes because fats in their entirety are too large to be absorbed by the cells in our intestinal wall. These digestive enzymes (that bear the beautiful name 'lipases') always break down fats in the same way: the head is separated from the arms. In short, the digestive enzymes decapitate the squid, so the head and arms float around free.

The arms, fatty acids, are small enough to be absorbed by the gut cells and enter the bloodstream, where they are absorbed by the cells in the body. The fatty acids are again recomposed in the cell, into a fully-fledged squid, or a fat. Fat cells are especially adept at this. They absorb large quantities of fatty acids (arms) from the blood and convert them into fats (squids).

Of great importance to our health is the understanding that *saturated* as well as *unsaturated* fats exist. The name 'saturated' or 'unsaturated' refers to the molecular structure of the fatty acids. In order not to make matters too complicated, we'll refer to fatty acids (arms) instead of fats (the arms and head). After all, it's the fatty acids that circulate in our bloodstream.

What are unsaturated and saturated fatty acids? Unsaturated fatty acids have one or more double bonds between their carbon atoms (C), which results in a kink. Saturated fatty acids don't have double bonds between the carbon atoms (C) and are completely straight.

The double bond between two carbon atoms causes these carbon atoms (C) to be strongly linked to each other. Normally, these two carbon atoms would have only one bond (a single line) with each other and a further bond with hydrogen atoms (H). But these hydrogen atoms (H) were removed by a chemical reaction and this enables the carbon atoms to bind more tightly together instead of with those fleeting hydrogen atoms.

(For those who really want to know the fine details: the bonds (lines) between the atoms consist of electrons, which form the molecular glue that holds the atoms together.)

Because there are two fewer hydrogen atoms at the level of the double bond, these fatty acids aren't completely saturated with hydrogen atoms (H); therefore they are *un*-saturated fatty acids. To put it another way, fatty acids are unsaturated because they're not completely saturated with hydrogen atoms because they have a double bond.

$$\begin{array}{ccccccccc} H & H & H & H & H & H & & H & H & H \\ | & | & | & | & | & | & & & | & | \\ C & C & C & C & C & C & = & C & C & C & -H \\ | & | & | & | & | & & & & | & | \\ H & H & H & H & H & & & H & H \end{array}$$

An unsaturated fatty acid has one (or more) double bonds between carbon atoms (C) (indicated by two lines above each other). Because of this, this fatty acid has a kink.

```
 H   H   H   H   H   H   H   H   H
 |   |   |   |   |   |   |   |   |
 C - C - C - C - C - C - C - C - C - H
 |   |   |   |   |   |   |   |   |
 H   H   H   H   H   H   H   H   H
```

A saturated fatty acid doesn't have a double bond between the carbon atoms and is thus made up of single bonds. Because of this, this fatty acid is straight.

Saturated fatty acids don't have double bonds. All the carbon atoms (C) are neatly bonded with hydrogen atoms (H). These fatty acids are therefore completely saturated with hydrogen atoms.

Why am I telling you all this? I am telling you because the fatty acid makes a kink at the level of a double bond. The carbon atoms on either side of the double bond are so strongly bonded together that the entire fatty acid bends. This molecular kink in the unsaturated fatty acids affects our health in the following way: it makes it difficult for these fatty acids to pile up next to one another and therefore, they are less likely to stick to one another. In fact, due to the kink, there is always some space between these molecules. Compare it to a wood pile. If all the logs have a kink, then the wood pile will be unstable: it will easily fall apart. But if the wood pile is made up of straight logs (as is the case with saturated fatty acids), then they can be packed tightly together and the wood pile will be much more stable.

Saturated (straight) fatty acids can therefore better position themselves next to one another, which enables them to clump together more easily. This way they can accumulate and adhere to the inner walls of our arteries, which can cause *atherosclerosis*. If these fatty build-ups in our arteries become large enough, they can clog the entire artery and we are talking again about a heart attack or a stroke. We've already seen that sugar plays the most important role in this process, and that saturated fatty acids are a less significant contributing factor to heart disease than previously thought. However some types of saturated fatty acids can also clump together more easily in our fat cells, which can make us gain weight at a faster rate. Some saturated fatty acids can also clump together anywhere in the body and cause inflammation, which can damage liver and brain cells.

In short, that ridiculously small molecular kink can make the difference between being healthy and sick, between good fats and bad fats that can adhere anywhere in the body.

SUMMARY

Saturated fatty acids (or fats):
- completely saturated with hydrogen atoms (H);
- the molecule is straight (there is no kink in it);
- these fatty acids can be packed tightly together, so that they can easily clump together anywhere in the body, and are also 'viscous' and harder in texture (like butter instead of oil).

Unsaturated fatty acids (or fats):
- the carbon chain isn't completely saturated with hydrogen atoms because there are one or more double bonds;
- the molecule has one or more kink in it;
- these molecules cannot come into complete contact with each other. They are more detached from each other, making a mixture of unsaturated fats more fluid (oil instead of butter).

The kink, or preferably the absence of one, can also explain why trans fatty acids (or trans fats) are unhealthy. Trans fatty acids have not yet been discussed. Trans fatty acids have a double bond just like healthy unsaturated fatty acids. But, unlike the ordinary unsaturated fatty acids with a double bond, the trans fatty acids don't have a kink, although they do have a double bond.

This occurs because, with trans fatty acids the hydrogen

$$\text{H}\ \text{H}\ \text{H} \quad\quad \text{H} \quad \text{H}\ \text{H}$$
$$\text{C-C-C=C-C-C}$$
(molecular structure diagrams)

Left, is an ordinary double bond. Because the hydrogen atoms (H) are located on the same level and on the same side of the double bond, the molecule can form a kink (there are no hydrogen atoms in the way). Right, is a trans fatty acid. The hydrogen atoms (H) are located on both sides of the double bond, resulting in a straight tail.

atoms (H) at the level of the two carbon atoms (C) are bound to opposite sides of the double bond (hence, the name 'trans', Latin for 'across'). With ordinary unsaturated fatty acids with a double bond, the hydrogen atoms (H) are located on the same side and form a kink. Most trans fatty acids that we eat aren't normally found in nature. In fact, Mother Nature has the habit of placing hydrogen atoms (H) on the same side of a double bond, as in the case of unsaturated fatty acids. The well-known and healthy omega-3-fatty acids are such fatty acids.

Because trans fatty acids are such strange fatty acids, the majority of proteins in our body can't process them very well. Trans fatty acids build up or produce foreign breakdown products that can be particularly unhealthy and because they have a straight structure like saturated fatty acids, they can easily clump together and cause atherosclerosis and micro-inflammations. Numerous studies indicate an influence of unnatural trans fatty acids in the development of cardiovascular disease. Trans fatty acids are formed when fatty acids are industrially prepared (think of biscuits, cakes, pastries, margarine, chips, croquettes and so on).

Playing with fats: margarine and trans fats

Interestingly, the structure of a fat or fatty acid at the molecular level (double bond or no double bond) determines how fatty acids react at the macroscopic level, like on the kitchen countertop. The more kinks in a fatty acid, the more liquid these fatty acids are. In fact, the kinks prevent the fatty acid chains from completely aligning them next to each other and sticking together. This prevents a solid, more set structure from forming. Just like crooked branches are also unable to form a stable, sturdy wood pile. Fatty acids with multiple double bonds (thus a lot of kinks) form the liquid oils such as olive oil or linseed oil.

However, it's not easy to spread oil on your sandwich. That's why the industry devised a process to convert vegetable unsaturated fatty acids with their kinks to saturated fatty acids without kinks, so these oils solidify and, for example, they can be spread onto bread. Such solidified oils are called margarine. The problem is, however, by creating saturated bonds from unsaturated bonds, trans fats are produced as a by-product. Thus, margarine originally contained a lot of trans fats that were particularly unhealthy to the blood vessels. Despite numerous publications in well-established medical journals, it took a few decades before the food industry recognised this problem. The margarine industry finally developed a process that produced margarine with considerably less trans fatty acids and yet, various margarines still contain unhealthy trans fats.

Margarine has the reputation of being healthier than butter, but that's merely an illusion; neither are advisable, margarine even less so than butter. Butter consists primarily of saturated animal fatty acids, while margarine is derived from unsaturated vegetable fatty acids that are chemically transformed into saturated fatty acids and the trans fats by-product.

Margarine	Butter
Consists primarily of unsaturated fatty acids from vegetable origin that are 'chemically modified' so that margarine contains saturated fatty acids and trans fatty acids.	Consists primarily of saturated fatty acids of animal origin.

Margarine manufacturers know very well that trans fatty acids and saturated fatty acids in their products aren't very healthy. So, to give their product a healthier image, some companies add phytosterols: a vegetable form of cholesterol that is supposed to lower cholesterol levels. Large advertising campaigns are then set up to convince the public how healthy their margarine is through TV ads with beautiful computer animations showing shimmering yellow cholesterol spheres and sporty, prominent countrymen who all laughingly spread their bread with margarine, or through nicely constructed websites where it is possible to assess your 'heart health'.

Scientists, however, aren't as convinced by phytosterols. There aren't any convincing studies that show phytosterols reduce the risk of cardiovascular disease. In fact, a recent overview article in the *European Heart Journal* even concludes the following, 'Study data suggests that phytosterols may cause adverse cardiovascular effects.'[48] In addition, the total cholesterol, that phytosterol should reduce, doesn't play a significant role in cardiovascular diseases, as we shall see. Nevertheless, margarine is being fortified with phytosterols and the marketing machine is doing the rest. The addition of dubious phytosterols in a substance that already contains unhealthy fats raises questions.

Trans fats are found mainly in fried foods and industrially prepared confectioneries such as tarts, biscuits, cakes and so on. It's probably best if you avoid this food as much as possible. Not only because of its content, but also because of what this food lacks. This junk food hardly contains any vitamins, minerals or healthy nutrients that the body needs to stay healthy.

SUMMARY

Very unhealthy fatty acids (or fats):
Trans fatty acids: are commonly found in margarine, fried foods and commercial confectioneries (tarts, biscuits, candy bars, cake).

Unhealthy fatty acids (or fats):
Saturated fatty acids: are commonly found mainly in animal products such as (red) meat, milk, butter, cheese, ice cream and also in (white) chocolate.
   Note: not all products with saturated fatty acids are unhealthy (dark chocolate can even be good for the heart and blood vessels).

Unsaturated fatty acids:
1. Poly-unsaturated fatty acids (have multiple double bonds and are therefore called *poly* unsaturated):
– omega-6-fatty acids: corn oil, sunflower oil, palm oil.
   Note: as we shall see, a dietary pattern that includes too much omega-6-fatty acids causes an increase in inflammation in the body.

Healthy fatty acids (or fats):
Unsaturated fatty acids:
1. Mono-unsaturated fatty acids (have one double bond and are therefore called mono-unsaturated): found in olive oil.
2. Poly-unsaturated fatty acids (have multiple double bonds and are therefore called *poly*-unsaturated):
– omega-3-fatty acids: commonly found in oily fish, walnuts, linseed, etc.

Omega-3-fatty acids: a hype in the media, but they do work

Fats don't always spell out doom and gloom. There are also healthy fatty acids; these are primarily unsaturated fatty acids. Scientists refer to these as mono-unsaturated or poly-unsaturated fatty acids, depending on the number of double bonds the unsaturated fatty acid has.

When the fatty acid consists of a single double bond, it is a mono-unsaturated fat. When it consists of multiple double bonds, it is referred to as a poly-unsaturated fat. Olive oil contains a lot of mono-unsaturated fatty acids. The renowned omega-3 fatty acids are poly-unsaturated fatty acids. Thus, omega-3 fatty acids contain various double bonds (for the chemists among us, the double bonds start from the third carbon atom when counted from the tail end). Omega-6-fatty acids contain double bonds (that start from the sixth carbon atom, when counted from the tail end). And omega-9-fatty acids contain double bonds (that start from the ninth carbon atom, when counted from the tail end). Lipidology is a very logical science.

Scientific research clearly shows that omega-3-fatty acids have numerous healthy affects on the body. The same can't be said for most of the omega-6- or omega-7-fatty acids; many of these fatty acids are unhealthy. For example, omega-6-fatty acids such as linoleic acid or arachidonic acid are pro-inflammatory. And yet despite this, countless dietary supplements containing omega-6-, omega-9- and even omega-7 fatty acids are now being sold.

But let us concentrate on the healthy omega-3-fatty acids. There are three types of omega-3-fatty acids:

Omega-3-fatty acids	Primary source
Eicosapentaenoic acid, EPA	oily fish
Docosahexaenoic acid, DHA	oily fish
Alpha-linolenic acid, ALA	nuts, linseed

A more visual presentation of omega-3- (and omega-6-) fatty acids:

alpha-linolenic acid (an omega-3 fatty acid)

eicosapentaenoic acid (EPA) (an omega-3 fatty acid)

docosahexaenoic acid (DHA) (an omega-3 fatty acid)

linoleic acid (an omega-6 fatty acid)

arachidonic acid (AA) (an omega-6 fatty acid)

A few omega-3- and omega-6-fatty acids. A carbon atom can be found at each angle. These carbon atoms are not depicted in accordance with standard scientific notation. The lines between these vertices or carbon atoms represent a bond. The 'omega' carbon atom is the utmost carbon atom to the right, thus on the end of the tail. The numbers 3 and 6 signify the third and sixth carbon atom, each time counting from the tail.

Omega-3-fatty acids have a proven influence on the following organs and systems:

- Heart and blood vessels: omega-3-fatty acids reduce the risk of cardiovascular diseases.

- The brain: omega-3-fatty acids reduce the risk of depression and other psychiatric disorders.
- The immune system: omega-3-fatty acids inhibit inflammation in the body.

Omega-3-fatty acids and the heart

Omega-3-fatty acids have a proven healthy influence on the heart and blood vessels.[49] Three large studies with more than 32,000 participants showed that omega-3-fatty acids, especially DHA and EPA, reduced the number of cardiovascular incidents by 19 to 45 percent.[50] It's noteworthy that this reduction was achieved despite the fact that every patient in this study had already been treated with all kinds of medications such as antihypertensives, statins, aspirin and beta blockers.

Another large study, in which more than 11,000 people took part and was published in *The Lancet*, revealed that the mortality rate fell by 45 percent in people who took omega-3-fatty acids (1000 mg/day). This study also showed that Vitamin E had no effect on cardiovascular diseases, although health gurus often cite this vitamin as cardioprotective.[51]

That is why official bodies are also pleading for the use of omega-3-fatty acids to combat cardiovascular disease. Researchers from the renowned Mayo Clinic in the USA promote the use of omega-3-fatty acids in order to prevent cardiovascular diseases in healthy people, as well as in those with a higher risk of cardiovascular diseases.[50] The American Heart Association (AHA) and the European Society for Cardiology (ESC) also recommend omega-3-fatty acids as an aid to preventing heart disease.[52-53] The American Food and Drug Administration (FDA) approve the use of omega-3-fatty acids to treat hypertriglyceridemia, a condition where the level of fat molecules (lipids) in the blood is elevated. When such official bodies recommend the use of omega-3-

fatty acids, the proof must have been raining down for many years.

It's interesting to note that in most studies, omega-3-fatty acids are administered orally to test subjects. But one person absorbs omega-3-fatty acids via the gut better than another. The quantity of fatty acids that is administered in pill form actually says very little about the quantity of omega-3-fatty acids that is actually present in the blood. It may be that one person could consume a lot of omega-3-fatty acids, but very little of this substance enters the blood. Because of this, this person is less protected against cardiovascular diseases and you could conclude that the administered omega-3-fatty acids have no effect. By determining the amount of omega-3-fatty acids in the blood, scientists can accurately assess the effect of this substance on heart and blood vessels. This has been researched and the results are remarkable.

A large study, in which more than 22,000 people took part and which lasted more than seventeen years, revealed that people who have a lot of omega-3-fatty acids in the

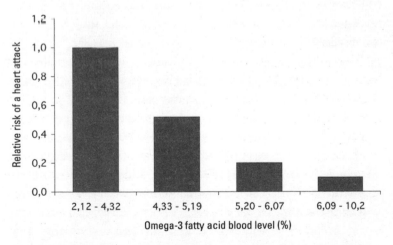

The more omega-3-fatty acids in the blood, the less risk of a heart attack. *Source: Blood levels of long-chain n-3 fatty acids and the risks of sudden death*, New England Journal of Medicine, 2002

blood are five times less (!) likely to suffer a heart attack.[54] These results were published in the prestigious medical journal, *The New England Journal of Medicine*. The conclusion from the researchers was that, 'omega-3-fatty acids found in fish are strongly associated with a reduced risk of sudden [cardiac] death.'

Interestingly, these results showed that people without a history of heart problems also benefitted from omega-3-fatty acids. The researchers even recommended those without heart problems to, 'increase the absorption of omega-3-fatty acids by eating more fish or taking nutritional supplements'.

However, there are some studies that don't demonstrate a beneficial effect of omega-3-fatty acids regarding the risk of getting a heart attack. It's a minority of the studies, but they do exist. In these studies some people with heart disease were worse off with omega-3-fatty acids. How is this possible?

As we shall shortly see, omega-3-fatty acids can also reduce the risk of cardiac arrhythmia, which is a good thing. Cardiac arrhythmia is caused by some of the heart muscle cells being very hyperexcitable and overactive. Omega-3-fatty acids reduce the irritability of these cells, which reduces the chance of over-stimulation of the heart (cardiac arrhythmia). But the hearts of patients with severe heart disease have great difficulty in keeping up their rhythm, because they have incurred too much damage due to previous heart attacks. In this case these overactive heart muscle cells may just keep the heart pumping sufficiently. If these people were to ingest omega-3-fatty acids, these heart muscle cells would no longer be sufficiently hyperactive, which could cause an aggravation of their heart problems.

Fortunately, this is very rare. In short, for the vast majority of the population, omega-3-fatty acids are wholesome, but people who suffer from serious heart failure are advised to consult their doctor before taking (high) doses of omega-3-fatty acids.

Countless studies show that omega-3-fatty acids have a protective effect on the cardio-vascular diseases. Even official bodies recommend the use of omega-3-fatty acids. Why then do we hear so little about it? Why do so many people think that it is not yet proven that omega-3-fatty acids are healthy for the heart, and why don't most doctors or cardiologists prescribe omega-3-fatty acids? There are several reasons for this.

To start with, omega-3-fatty acids can't be patented. Omega-3-fatty acids exist naturally in fish and nuts. They are not synthetic substances made in a laboratory that can be patented like most medicines. That is why it is useless for pharmaceutical companies to spend tens of millions of dollars on an advertising campaign for a product that anyone can introduce onto the market. So, the advertising budget available to omega-3-manufacturers is significantly smaller than that of the pharmaceutical industry. The advertising budgets of the pharmaceutical industry are astronomical, certainly with regard to the huge profits that can be made when a medicine is introduced onto the market after many years of research.

In 2003, the pharmaceutical company Pfizer made a profit of 9.2 billion dollars solely on its cholesterol-reducing medicine, Lipitor (a statin). With such sales figures you can make the odd advertisement. That's a reason why medical journals aren't filled with adverts for omega-3-fatty acids, but with patented medicines, and doctors aren't visited by representatives with samples of omega-3-fatty acids in their bags, but samples of statins.

Another reason why omega-3-fatty acids are prescribed in such a limited way is that there are few published scientific studies on omega-3-fatty acids. Ninety percent of all medical studies are funded by the pharmaceutical industry. Scientific studies are expensive and a pharmaceutical company naturally prefers to fund research into its own medicines.

A drug company is hardly going to fork out five million dollars for a study into fish oil capsules that everyone can introduce onto the market. Luckily, there is research carried out into non-patented medicines or dietary supplements, usually by universities or the government. But these studies are in the minority compared to studies into regular medicines. And ignorance breeds contempt.

Finally, the pharmaceutical industry has its own alternatives for omega-3-fatty acids, namely its own unique medicines for treating cardiovascular diseases: the fibrates and the renowned statins. Fibrates are substances that regulate the metabolism of fat. Fibrates lower the level of bad cholesterol (LDL) in the blood and raise the level of good cholesterol (HDL). They also lower the level of lipids in the blood (the *triglycerides*). Sounds good in theory. But in practice, fibrates have not had any proven affect on the mortality rate.[10] Whether you are taking fibrates or not: you do not live a day longer! In addition, fibrates have numerous side effects; the most common are muscle pains, erectile dysfunction, liver and gallbladder disorders and an increase in *homocysteine* levels. The latter is interesting. Homocysteine is ironically enough one of the risk factors for cardiovascular disease. It's possible that the health benefit gained from lowered levels of bad cholesterol and lipids is offset by the increased homocysteine. These types of interactions aren't surprising. Fibrates are basically medicines, and lots of medicines inhibit a protein that has multiple functions in the body and this can lead to side effects. Fibrates, like most medicines, are foreign to the body (xenobiotic), while omega-3-fatty acids are natural substances found in humans and animals.

In addition to fibrates, statins can also be prescribed in cases of cardiovascular disease. Unlike fibrates, statins can reduce the mortality rate, thereby improving survival rates, especially in heart patients. However, statins also have side effects such as *rhabdomyolysis* (breakdown of muscle), liver disorders, erectile dysfunction or sleep and memory disorders. It's often claimed that these side effects are rare, but that depends on personal interpretation. When doctors speak of side effects, they are referring in the first instance to grave side effects such as massive muscle breakdown that may or may not lead to renal failure, a severe allergic reaction or acute liver failure. Of course, these types of cases occur very rarely with statins. But statins inhibit cholesterol production. Cholesterol is often portrayed as the great evil-doer; but cholesterol is an important substance in the body and it fulfils numerous crucial functions.

Cholesterol is, for example, an important component of the walls of our cells, especially for muscle cells and nerve cells. Cholesterol embeds itself in the walls of these cells, making them fluid. The walls are thus less rigid. When statins are administered, less cholesterol is produced and the membranes in muscle cells and nerve cells become harder which weakens them. This can induce muscle pain and loss of strength. This muscle breakdown is seldom so serious that massive muscle deterioration ensues and patients find themselves at the hospital emergency department, but a lot of people who take statins have typical 'ailments of old age' such as painful muscles, muscle fatigue, nerve pain or memory complaints and difficulty concentrating.

With regard to muscle pain, it can be said that the more the patients move, the greater the damage to muscle cells by statins. Although studies show that the risk of serious side effects from statins is less than 0.1 percent, 80 percent of young athletes who take statins and who continue to do sport have muscle complaints.[55] You might be wondering why young people who do sport have to take statins, but this was a study carried out on sportspersons who suffered from 'familial hypercholesterolaemia'. This is a genetic disorder and these people have excess cholesterol in their body. So statins should ideally work most effectively with them and produce the fewest side effects (they have abnormally high levels of cholesterol, thus reduction is really necessary) and yet these young athletes suffered from muscle complaints brought on by the use of statins. According to Professor Helmut Sinzinger, the author of the study, more than 25 percent of people who are healthy (and thus not suffering from familial hypercholesterolaemia) could develop muscle pain from intense sporting activities. That is a different set of figures to the 0.1 percent risk of (serious!) side effects by the use of statin.

Forgetfulness and concentration disorders can occur from the use of statins because cholesterol is also necessary for fluidity of the brain cell walls. When this becomes less flexible it can lead to memory problems. Nerve pain may also occur. A study published in *Neurology* showed that people who took statins for at least two years were more susceptible to nerve pain (polyneuropathy).[56]

But statins can reduce the risk of a heart attack. And omega-3-fatty acids can also reduce the risk of a heart attack. Which of the two would be best: the statins or the omega-3-fatty acids?

A large analysis of 97 scientific studies was undertaken in which a total of 140,000 patients took part. In this analysis, researchers sought to verify which of the two is the better: the statins or the omega-3-fatty acids. The result was that, from all known interventions to reduce general and cardiovascular death, the omega-3-fatty acids were better. This study was published in the *Archives of Internal Medicine*, one of the foremost medical journals.[57] The following is a list of the different interventions that, according to this study, affected the mortality rate.

*   *   *

Intervention	Mortality (1 no effect; the lower, the better)
Fibrates	1
Consultation with a dietician	0.97
Statins	0.87
Omega-3-fish oil	0.77

Source: Effect of different antilipidemic agents and diets on mortality: A systematic review. Archives of Internal Medicine, 2005

It seems clear from this study that omega-3-fatty acids work best with regard to general mortality as well as cardio-vascular death. Omega-3-fatty acids are therefore even more effective than the statins that are prescribed. It is note-worthy that consultation with a dietician has little influence on the mortality rate. After all, the dieticians in this study rely primarily on the old food pyramid and low-fat diets to prevent cardiovascular diseases.

However, statins can be useful within the scope of prevention if patients have already experienced a heart attack or minor stroke, or have major clogging of the blood vessels in the heart or brain. The problem though is that nowa-days statins are simply prescribed far too often, for instance, to people who just suffer from elevated cholesterol levels, without even trying other treatments, or trying to tackle the cause of these high cholesterol levels. Big pharma makes sure a steady stream of scientific articles is produced that extol the virtues of statins, with the result that some medical doctors even propose adding statins to drinking water or including a package of statins along with every hamburger, so that you could sprinkle statins on your Big Mac. How crazy is that? Shouldn't we deal with the hamburger first,

instead of sprinkling the dire consequences away with medication?

Why can omega-3-fatty acids be heart-protective? What is
the mechanism behind that? Just as with cholesterol, omega-
3-fatty acids get embedded into the walls of cells that make
up our heart and blood vessels. Because of this, the cell walls
become more fluid, making it easier for proteins to float
around in the cell walls and interact with each other.[58]

The walls of blood vessels become healthier because of
this. Healthy blood vessel walls ensure that fewer clumps are
formed in the blood vessels and so the risk of a heart attack
is reduced. Omega-3-fatty acids are not only embedded into
the blood vessel cells but also into the walls of *thrombocytes*.
Thrombocytes are small, flat blood cells that clump together
to form clots. They play an important role by clotting when
we are injured, but they can also spontaneously clot in our

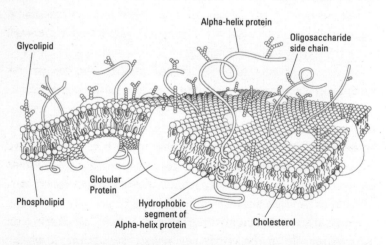

A piece of a cell wall consists of a sea of fats in which proteins
(the large spherical structures) float around like buoys. A cell is
actually a miniscule soap bubble because the wall of a soap
bubble is also composed of fats (the branchlike protuberances
are sugar chains). *Source: Dana Burns*, Scientific American,
1985

bloodstream. When the walls of the thrombocytes contain enough omega-3-fatty acids, then the thrombocytes are less likely to clump together.[59] The omega-3-fatty acids can also embed in the heart muscle cells. When the heart muscle cells contain a lot of omega-3-fatty acids the heart is less susceptible to heart rhythm disorders or arrhythmia.[60] Arrhythmias are involuntary and chaotic contractions of the heart that commonly occur in the elderly. More than a quarter of all westerners between 50 and 79 years old will experience atrial fibrillation, where the upper chamber of the heart contracts too fast. Such rhythm disorders are caused by charged particles that pass uncontrolled through the gates in the cell walls and enter the heart muscle cells. Research in which 160 patients took part showed that patients who received omega-3-fatty acids before they underwent heart surgery had 54 percent less risk of experiencing atrial fibrillation.[61]

But omega-3-fatty acids don't just embed in the walls of the heart and blood vessels. They also embed in the wall of our brain cells. In this way, omega-3-fatty acids can have a preventive and sometimes even a curative effect on various brain disorders.

Omega-3-fatty acids and the brain

Seventy-seven percent of our brain consists of water. If we removed all of that water, then of the dry-weight that remains 60 percent is fats. Thus, our brain consists primarily of fats. The brain is a 'fat organ'.

Thus, it's not illogical to assume that fatty acids play an important role in the construction and maintenance of the brain. Certainly with regard to essential fatty acids such as omega-3-fatty acids which the body can't produce and of which most westerners ingest very little in their diet. The first important hint highlighting this was given in an article published in *The Lancet*. In it is a chart depicting a link between the consumption of fish and the prevention of depression in different countries.

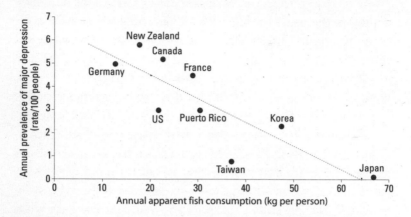

Serious depression occurs much more in countries where little fish is eaten. *Source: Fish consumption and major depression,* The Lancet, *1998*

Depression is a lot more common in countries where little fish is eaten, such as Germany or Canada, than in a country like Japan, where a lot more fish is consumed. Researchers speak of a factor 60. In short, depression, recognised as a common condition that will soon take the number two spot behind cardiovascular disease in the West, occurs at a much lesser degree in some countries. Interestingly, the researchers also determined a similar link between the consumption of fish and cardiovascular diseases.

Numerous studies, published in leading psychiatric journals such as the *Archives of General Psychiatry*, show that omega-3-fatty acids can indeed protect against depression and can, for example, reduce depressive episodes in manic-depressive people. A study of manic-depressives showed that after four months of taking omega-3-fatty acids, approximately 90 percent of the patients did well (they didn't relapse into depression), while of those who didn't receive the omega-3-fatty acids, but just the usual medication, only 40 percent didn't relapse.[62]

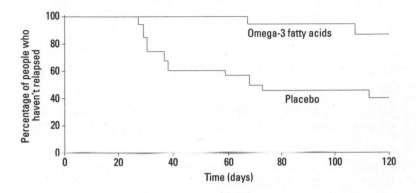

Test subjects who receive omega-3-fatty acids have much less chance of relapsing into a depression. *Source: Omega 3 fatty acids in bipolar disorder: a preliminary double-blind, placebo-controlled trial.* Archives of General Psychiatry, *1999*

In another study, people were given 2 grams of omega-3-fatty acids per day. After a month, their score on the Hamilton Depression Scale (a scale that measures the seriousness of depression) dropped by half from 24 to 11.6. While the score from the control group who received placebos (a fake medication) remained the same.[63] Another study published in 2010, and in which 432 patients took part, once again confirmed that omega-3-fatty acids 'were superior' when compared to placebos.[64]

There are even studies that show omega-3-fatty acids can help against psychotic disorders. Psychoses are often serious psychiatric syndromes where the brain can become damaged and which are treated with strong medication. A psychotic disorder occurs, for example, in people with schizophrenia. Someone with a psychosis suffers from delusions and hallucinations such as hearing voices in their head, seeing things that are not there or having delusional thoughts, such as the feeling of being followed or the idea that someone has implanted a transmitter in their head.

In a study, again published in *Archives of General Psychiatry*, people with a high risk of developing a psychotic disorder were given omega-3-fatty acids. After 12 months only 4.9 percent of those who took omega-3-fatty acids suffered a psychotic disorder, compared to the 27.5 percent of those from the group who didn't receive omega-3-fatty acids. In short, people who hadn't taken omega-3-fatty acids were five times more likely to become psychotic.[65]

Omega-3-fatty acids can also play a role in postnatal depression. Postnatal depression occurs when a mother develops a depression in the first weeks after the birth of her child. It is often claimed that this depression is a result of a change in hormone levels. Yet another explanation is that omega-3-fatty acids play a role. During the growth of the foetus, omega-3-fatty acids are important for the development of the foetal brain. Growing a baby's brain requires a lot of omega-3-fatty acids and when there isn't enough available, the child will extract what it needs from its mother. Because of this, the mother becomes deficient in omega-3-fatty acids and more susceptible to depression.[66]

The fact that omega-3-fatty acids play an important role in the growth and development of the foetal brain has been shown in a number of studies, among them one published in *The Lancet*. Children whose mothers ate sufficient fish during the pregnancy had a higher IQ than children whose mothers consumed less fish. Verbal and cognitive intelligence, and motor skills improved considerably when mothers consumed sufficient omega-3-fatty acids during pregnancy; especially through eating fish. Mothers who had eaten sufficient fish had a fifty percent reduction in the chance that their child would have a lower verbal IQ.[67]

Ironically enough, pregnant women are actually often discouraged from eating oily fish. The reason being that oily fish could be polluted with mercury and other toxic substances, which could have a negative affect on the unborn child. The researchers however concluded:

> 'We recorded beneficial effects on child development with maternal seafood intakes of more than 340 g per week, suggesting that advice to limit seafood consumption could actually be detrimental. These results show that risks from the loss of nutrients were greater than the risks of harm from exposure to trace contaminants in 340 g seafood eaten weekly.' *(Maternal seafood consumption in pregnancy and neurodevelopmental outcomes in childhood (ALSPAC study)*, The Lancet, 2007)

Omega-3-fatty acids do not only play a role in the development of the young brain, but also in the maintenance of the old brain.

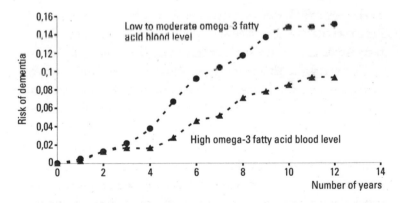

People with low levels of omega-3 in the blood (spheres) have a greater chance of developing dementia compared to those who have high levels of omega-3-fatty acid in the blood (triangles). *Source: Plasma phosphatidylcholine docosahexaenoic acid content and risk of dementia and Alzheimer's disease: the Framingham Heart Study,* Archives of Neurology, 2006

In a study published in *Archives of Neurology* in which 899 people were followed for a period of more than nine years, it was found that those with high levels of the omega-3-fatty acid, docosahexaenoic acid (DHA), in the blood had a fifty percent reduction in the risk of developing dementia, compared to people with lower DHA-levels.[68] The researchers also observed that eating oily fish twice a week was sufficient to ensure this association between fish oil and this fifty percent reduction in the risk of dementia: 'In our study, a 50% reduction in the risk of Alzheimer disease was associated with the consumption of more than 2 servings of fish per week.'

A lot of studies show that omega-3-fatty acids have a positive effect on the brain and that they can reduce the risk of, or even cure, numerous psychiatric disorders such as depression. In a large analysis of dozens of scientific studies, carried out by eleven specialists, the following was concluded:

The preponderance of epidemiologic and tissue compositional studies supports a protective effect of omega-3 essential-fatty acid intake, particularly eicosapentaenoic acid (EPA) and docosahexaenoic acid (DHA), in mood disorders. *(Omega-3-fatty acids: evidence basis for treatment and future research in psychiatry,* Journal of Clinical Psychiatry, 2006)

The American Psychiatric Association (APA) recommends the use of omega-3-fatty acids too, for depression and psychotic disorders such as schizophrenia. The APA is the professional organisation of psychiatrists in the USA and is one of the most important organisations in psychiatry, which, among other things, also publishes the well-known *Diagnostic and Statistical Manual of Mental Disorders* book, also known as the 'bible' of the psychiatry.[69]

How is it possible that fatty acids can affect a range of different and diverse mental disorders?

In the first place, it is because the walls of our brain cells consist primarily of fats, as we have already seen. Omega-3-fatty acids embed themselves into the walls of our brain cells so that they become more fluid. Because of this, the proteins floating around in them can communicate better, especially at the level of the synapses. Synapses are places where two nerve cells make contact with each other and exchange signals. The more fluid the synapses, the better and easier it is for nerve cells to exchange signals: the brain becomes more powerful as it were and therefore, more resistant to depression and mental illness.

In addition, omega-3-fatty acids also reduce *micro-inflammation* in the brain. Micro-inflammation refers to inflammation at a cellular level, whereby certain cell components, such as cell walls, become damaged and inhibit the optimal working of the nerve cell. In addition to these biochemical reasons, there are evolutionary reasons why omega-3-fatty acids are so important for the brain. Omega-3-fatty acids are components of fats and quite possibly played an important role in the evolution of man, especially in the growth of his brain. It's possible that certain gene mutations that play a role in the metabolism of fat are an important reason man became smarter,[70] the reasoning being that a large part of the human brain consists of fats. If our ancestors could better incorporate fats into their brains because of a mutation then they were able to develop larger and more powerful brains. These mutations, which changed the fat metabolism entirely, inevitably caused changes in other parts of the body too. The fats in the body got distributed differently, which would explain the difference between humans and great apes with regard to fat distribution in the body. Great apes such as bonobos or chimpanzees have very little subcutaneous fat compared to people. Moreover, bonobos also have less fat in the breasts and buttocks. Compared to the average bonobo female, human females have gigantic breasts and buttocks (composed of adipose tissue). In short, a different fat distribution in the body (more fat for the brain, breasts, buttocks, and under the skin) could have ensured that people became more intelligent compared to the great apes. A disadvantage however of a large and powerful brain could well be that it's more susceptible to mental illnesses like schizophrenia. According to some scientists, schizophrenia is the toll we have to pay for our increased intelligence and creativity. One percent of the population is schizophrenic or psychotic, a significant percentage.

Some people may wonder how it's possible that the omega-3-fatty acids could play such a prominent role in human evolution; the arid African savannah plains where we evolved didn't particularly swarm with omega-3-oil-rich fish. On an arid savannah plain you not only find few fish, but also very little other food. The threat of famine was a great problem for our distant ancestors, even more so because the savannah was increasingly plagued by periods of drought. That process began millions of years ago, caused by the climate change as a result of Africa splitting in two. Indeed, even now Africa is continuing to break in two and East Africa is in the process of tearing away from West Africa as a result of the movement of continental plates, floating on a sea of magma. The result of this schism is the Great Rift Valley, an enormous fissure, thousands of kilometres in length, which runs from the north of Africa to the south. A few of Africa's largest lakes and waterfalls are the result of the still widening rift.

This process started millions of years ago. The climate remained warm and humid to the west of the rift with plenty of jungle – that was where the ancestors of the great apes such as the bonobo and chimpanzee were found. To the east of the rift the climate was much drier, forming dry savannah plains (as is still the case) and our ancestors evolved in this dry landscape. Unlike the ancestors of the bonobos and chimpanzees who lived in the lush jungle west of the Rift Valley, our ancestors had to get smarter and more skilled to find food on the dry savannah landscape. Because of this, they developed a larger brain and the ability to walk upright. In short, these geological and climatic changes around the Great Rift Valley made it possible for our species to evolve and they also explain why man differs so much from his closest relatives such as the bonobos and the chimpanzees (who didn't have to evolve as much and become smarter because they had enough shelter and food in their jungle).

However, at some point in time on the east of the rift it became so dry that our ancestors could barely find enough food on the dry savannah plains inland and they took refuge on the coast. The humid air ensured a less dry climate and moreover our ancestors could feed on fish, crustaceans and shellfish from the coast. This meant they could ingest the necessary omega-3-fatty acids that would play such an important role in the further growth and development of the brain. We are all descended from these ancestors who, during these extended periods of drought, lived in caves on the African coast.[71]

So there aren't just biochemical reasons, but also evolutionary reasons too why omega-3-fatty acids are so important for our brain.

Omega-3-fatty acids and the immune system

The human body is in a constant state of inflammation at some point or other. Micro-inflammations play a role in atherosclerosis, cancer and even dementia. But inflammation on a larger scale can cause inflammatory diseases. These diseases are caused by our immune system, which sometimes becomes overactive or attacks the body's own tissue which results in numerous inflammatory diseases. Rheumatism is a disease where white blood cells attack the joints; asthma is a disease where white blood cells are stimulated too quickly and attack the bronchi; eczema is a disease that occurs in the skin; Crohn's disease is a disease where the immune system in the intestines is overactive; and with lesser known diseases such as lupus, scleroderma or polymyositis, the immune system considers its own muscle and connective tissue as foreign.

Numerous studies show that omega-3 fatty acids reduce inflammation and symptoms in immune diseases such as rheumatism[72-73], asthma[74] and Crohn's disease.[75] Administration is important here. With rheumatism, high doses of omega-3-fatty acids are needed (think of a few grams per day) and a visible effect will be seen only after three months. To generate an effect in Crohn's disease, administered pills must be enteric coated so they don't immediately dissolve in the stomach, but in the intestine so that they can discharge their content where the inflammation occurs.

The reason why omega-3 fatty acids can inhibit inflammation, is because they act on an important protein in the body called COX (cyclooxygenase). This is a protein that creates substances that cause inflammation. Aspirin, the well-known anti-inflammatory medicine, acts on this protein by inhibiting and blocking it, preventing it from making pro-inflammatory substances.

However, the cox-protein not only makes pro-inflammatory substances but also anti-inflammatory compounds. It makes anti-inflammatory compounds from omega-3-fatty acids while pro-inflammatory substances are made from omega-6-fatty acids. That is why it is so important to have a good balance between omega-3- and omega-6-fatty acids. However, the modern western diet contains much more pro-inflammatory omega-6-fatty acids (primarily found in meat, margarine and oils such as corn and sunflower oil) than the anti-inflammatory omega-3-fatty acids and this can play a role in numerous cardiovascular diseases and immune disorders.

How much omega-3?

To obtain a protective effect for heart and blood vessels the recommended dose is at least 500 milligrams (mg) omega-3-fatty acids per day in the form of EPA (eicosapentaenoic acid) and DHA (docosahexaenoic acid). For people who are at risk from cardiovascular disease (for example due to being overweight or because they have family members with heart problems or have already suffered a minor heart attack) at least 1 gram (1000 mg) per day is recommended. The same quantities apply for depressions. For prevention, 500 mg of omega-3-fatty acids is sufficient. To really treat depression, 1, 2 or more grams of omega-3-fatty acids is used in research. It's not necessary to take omega-3-fatty acids every day; these fatty acids remain present for weeks in the membranes of our cells.

Different research shows that with cardiovascular disease, DHA plays a particularly favourable role, while in treating depression and other mental disorders, EPA is the better choice. However, I always recommend using a combination of DHA and EPA because in this way the heart, as well as the brain, will be protected and they strengthen each other's benefits. The recommended amount of 500 mg of omega-3-fatty acids per day is the sum of EPA and DHA. For example, if there are 200 mg EPA and 300 mg DHA in a capsule, then you will have 500 mg of omega-3-fatty acids per day, exactly enough.

EPA and DHA can be found primarily in oily fish, which means oil-rich fish like salmon, mackerel, herring, anchovies and sardines. I call this, 'the fatty five'. It is often claimed, with regard to salmon, that wild salmon contains more omega-3-fatty acids than farmed salmon. That is not the case. Farmed salmon contains more omega-3 than wild salmon[76] because farmed salmon is well fed and doesn't have to swim around looking for food.

Eating oily fish just twice a week is the same as taking one 500 mg capsule of omega-3-fatty acids every day. So you don't necessarily need to take dietary supplements: they cost a lot of money and it is cheaper and healthier to ingest the omega-3-fatty acids by eating oily fish. Moreover, fish oil in capsules is often oxidised, so the fatty acids are rancid and do not work so well. Sometimes, people who take omega-3-supplements can develop (minor) liver disorders because of this. That is why it is better to store omega-3-supplements in the fridge. However, by the time they arrive in your fridge, they have already been stored on a warm shelf in a warehouse and a shop for several months. Moreover, an oily fish on your dinner plate contains many other healthy substances than just omega-3-fatty acids, substances that are not always found in food supplements such as furan fatty acids found in fish that are also partly responsible for the protective effect on heart and blood vessels.[77] Don't get me wrong, supplements are not bad, but oily fish on your dinner plate is even better.

It is possible that some fish oil capsules can be heavy on the stomach and some people feel slightly nauseous or complain about fish-breath. This can easily be solved by taking the capsules in the morning before breakfast (so that the capsule is covered by the food) or, by storing the fish oil in the fridge or, by buying *enteric-coated* capsules. These capsules are enveloped in a small layer that protects the capsules from the effects of stomach acid so they only burst in the intestines and feel less heavy on the stomach.

And what about the accumulation of mercury in fish? Salmon, mackerel and other oil-rich fish swim around in our polluted oceans where mercury could accumulate in their bodies.

That all sounds plausible, but research shows that the amount of mercury in oily fish is very low, whether the fish swims around free or was farmed. Researchers also concluded that fish oil supplements 'contain negligible amounts of mercury'.[78-79] Of course, other substances other than mercury can accumulate in fish. It is also true that, from a toxicological point of view, the larger the fish, the more toxic substances can accumulate in the fish because larger fish eat smaller fish, which also contain a quantity of toxic substances. These toxic substances are not excreted by the larger fish, but accumulate in their fatty tissue. In short, an anchovy would have less toxic substances per gram of tissue than tuna or a large swordfish. The least accumulation occurs in very small marine organisms such as algae and crabs. Some omega-3-supplements are therefore acquired from these organisms. For those unwilling to take any chances, they can eat small oily fish species such as an anchovy or mackerel or take algae-based omega-3 supplements.

SUMMARY

Despite the fact that westerners eat increasingly **fewer fats**, they are becoming increasingly **fatter**.

People who follow a low-fat diet lose fifty percent less weight than those who follow a low-sugar diet. To a lesser degree fats, but primarily **sugars** play a role in cardiovascular disease.

There are, however, **healthy** and **unhealthy** fats.

The very unhealthy fats (or fatty acids) are:
- **trans fatty acids**: commonly found in industrially prepared foods, such as margarine, fast food and ready meals, fried foods such as chips and croquettes and pastries such as biscuits, cakes, tarts, etc.

Continued overleaf

Less healthy fats (or fatty acids) are:
- **saturated fatty acids**: in animal products high in fat such as meat, milk, cheese and in vegetable products such as coconut milk, palm oil, chocolate, etc. Note: not all products with saturated fatty acids are unhealthy (for example dark chocolate)
- **omega-6-fatty acids**: in meat and in vegetable oils such as sunflower oil, palm oil, corn oil etc.

Healthy fats are:
- **omega-3-fatty acids**: such as in oily fish, nuts and linseed;
- **mono-unsaturated fatty acids**: such as in olive oil.

Note: omega-6- as well as omega-3-fatty acids are poly-unsaturated fatty acids.

Omega-3-fatty acids:
- according to many large studies, reduce the risk of a **heart attack**;
- according to a study with 140,000 patients, **work better than statins** with regard to cardiovascular disease;
- can dramatically reduce the likelihood of relapse in **depression, psychoses and manic-depressives** and play a role in the development of the brain of the foetus;
- are recommended by **official bodies** such as the American Psychiatric Association (APA), the American Heart Association (AHA) and the European Society for Cardiology (ESC);
- inhibit **inflammation** and can provide a supporting role by auto-immune diseases.

4

The food hourglass

'Know thyself' is inscribed above the famous Greek temple at Delphi. This motto also applies to our food. 'Know thy food' or know what you eat. In the previous chapter we discussed the three major calorie groups (sugars, proteins and fats). Now let's examine the different foods that make up the food hourglass and why these products are so healthy or unhealthy.

Step 1: Beverages

Soft drinks, milk, (drinking) yoghurt,
non-freshly squeezed fruit juices.

replace with

Water, green tea, white tea, ginger tea, red wine,
freshly squeezed fruit juice, vegetable milk (including:
soya milk, almond milk, rice milk), coffee.

Water
Kidneys are amazing, complex machines. Each kidney consists of one million nephrons and each nephron is a sort of mini-kidney, the basic unit of the kidney that actually performs the filtering function. Each nephron consists of a

tangle of blood vessels inside a 'funnel'. Fluids (containing toxins and waste products) trickle out of this tangle of blood vessels from the blood into the funnel and discharged into the ureter, which sends urine towards the bladder.

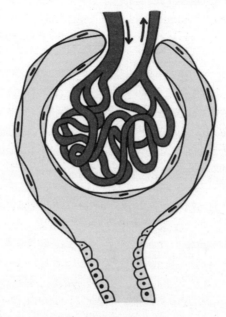

Fluid is 'squeezed' out of the tangle of blood vessels into a funnel. Most of this fluid from the bloodstream is reabsorbed further down the funnel so that it re-enters the bloodstream. The remaining fluid flows through to the ureter (which discharges into the bladder).

Approximately 180 litres of fluid are squeezed out of the bloodstream into two million nephron funnels every day. Most of the 180 litres is reabsorbed and re-enters the bloodstream. This means that every day the five litres of blood in your body are filtered approximately 36 times. Approximately 1.5 litres of the 180 litres is retained in the nephrons, and this is the one and a half litres of urine that we excrete each day. This means that our kidneys are working hard all the time which is why they are very sensitive to the ageing process.

From the age of thirty, our kidney capacity will drop by 10 percent every ten years. This means that most seventy- and eighty-year-olds retain just half of their kidney functions. In most cases, it is actually less than half due to unhealthy lifestyles. Three major causes of kidney damage, leading to faster deterioration are:

- drinking insufficient water;
- eating too much protein-rich food;
  use of medicines, including high doses of aspirin, some blood pressure medication or diuretics.

Drinking enough water is vital for the kidneys. The more water you drink, the more water passes through your kidneys. The passage of sufficient water ensures that the thousands of kilometres of vessels in our kidneys can expand (because they are filled with more liquid), so that the kidneys are less likely to be damaged. Extra water also ensures that the kidneys are able to filter large amounts of blood. Thus, water lubricates the kidneys in the same way that oil lubricates a machine. The importance of drinking enough water can be clearly seen whenever there is a heat wave. Older people in particular who don't drink enough on a hot day run the risk of kidney damage and even kidney failure. Kidney failure is an all too frequent occurrence in nursing homes on a hot summer's day.

Drinking enough water is important not only for the kidneys, but also for the heart and blood vessels. According to the *Adventist Health Study,* in which more than 20,000 test subjects took part, people who drink little water (two glasses a day or less) have twice as much chance of getting a heart attack than people who drink at least five glasses of water each day.[80] Doctor Gary Fraser, cardiologist and statistician, recommends drinking five or six glasses of water a day. Taxi drivers who work at night can significantly diminish their chances of getting a heart attack by drinking more water.[81]

It is important to drink enough fluids every day, principally water. A human being needs two litres of fluid daily, in the form of drinks (there is also extra fluid in food). The earlier you start taking in fluids during the day, the better. This is why the renowned physician Hiromi Shinya drinks two glasses of water every day on waking, fifteen minutes before breakfast. This ensures that the body starts absorbing liquid right from the start of the day and that the kidneys can build up their filtering capacity. This also wakes up the digestive tract. Not only that, drinking one or two glasses of water before breakfast will ensure that, after about a quarter of an hour, your appetite will increase so you can eat a good breakfast, which, as we will see later on, is very important. So, after drinking your water, do something else for fifteen minutes before eating breakfast: have a shower, get dressed, have a shave, do your make up, read the paper or prepare lunch.

As well as drinking too little, an excess of protein is also bad for the kidneys. In the previous chapter we saw that meat, a rich source of proteins, can overburden the kidneys. However, it is not just an excess of proteins, but also many medicines that can damage the kidneys, for example, antibiotics or acyclovir taken by mouth (for cold sores). These chemicals are excreted by the kidneys but they are especially damaging and toxic for them. Painkillers including diclofenac, ibuprofen, naproxen, (high doses of) aspirin and certain medicines for lowering blood pressure cause constriction in the blood vessels in the kidneys, restricting blood flow, which can also damage the kidneys.

The result is that many long-term medicines including

painkillers and medication for hypertension, can damage the kidneys.

This means that you must be careful with medicines as well as proteins and above all, drink plenty of water. There are, of course, many other healthy drinks as well as water that can, in their own way, have a positive effect on the body.

Green tea, white tea and ginger tea

Countless studies have shown that green and white teas have a particularly healthy influence on the body. Green tea can reduce the risk of cancer and can affect our metabolism in such a way that we can lose weight. White tea can reduce the onset of wrinkles because it inhibits *proteases*. Allow me to explain.

Proteases are proteins that break down the collagen and elastin in our skin; collagens and elastins are also proteins. Collagens are long strings of proteins that keep the skin supple; elastins are springy proteins that give the skin its elasticity. When these proteins are broken down by the proteases the skin becomes less elastic and this is one way wrinkles are formed. Eating too much sugar will also speed up the formation of wrinkles, as we have already seen in Chapter 3.

If you look at the diagram on the next page, you can clearly see that white tea is the most powerful inhibitor of the proteases that slowly, but surely, break down the skin's components as the years pass by. White tea inhibits both collagenases and elastases (the names of the proteins that break down collagens and elastins respectively). The substance known as EGCG, one of the most important constituents of green tea, is just as good as white tea at inhibiting elastinase, but less effective at inhibiting collagenase. This means that green tea also has a positive influence on the ageing process of the skin but probably less than that of white tea.[82]

In recent years, green tea has been the subject of more

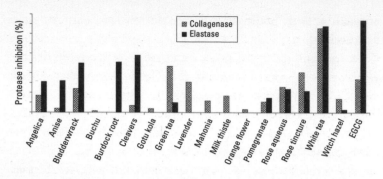

This diagram shows the extent to which certain food products such as white tea or EGCG (contained in green tea) inhibit proteases, proteins that break down the skin components thus causing wrinkles. *Source: Anti-collagenase, anti-elastase and anti-oxidant activities of extracts from 21 plants,* Biomedcentral, 2009

studies than white tea. We are frequently told that green tea is so healthy because it contains antioxidants. However this is not the reason why green tea has such positive effects on the body. First let's study this antioxidant myth in more detail.

We read and hear in the popular press and particularly from food supplement manufacturers that antioxidants are healthy. Many scientists (who are the *real* specialists in antioxidant research) do not agree with this at all. The general public is told that antioxidants are healthy because they neutralise *free radicals*. Free radicals are small molecules that are produced throughout the body as by-products of our metabolism. They react with proteins, DNA fragments or the cell walls, damaging them. Antioxidants clear away these free radicals by reacting with them.

Antioxidants are the bodyguards that take the bullet. They do this first of all by reacting with the free radicals so that the valuable cell components such as DNA or proteins are protected from attack by the free radicals. However, if the antioxidants in green tea are as effective at combating the free radicals as is claimed, we have a problem: we actually need free radicals. We need them because they set off the alarms of our cells. When our cells detect free radicals, they will then start producing all kinds of proteins to protect themselves. These proteins are much better able to combat the free radicals than the antioxidants that you ingest via food or food supplements. This means that when we ingest large amounts of antioxidants our cells will not be alarmed and this, in turn, will cancel our cells' defence mechanisms, which protect us from many, more dangerous free radicals and toxic materials. The weakened defence mechanisms will allow the toxic materials and free radicals to do their nefarious work, which in the long term will damage our health. This also explains why (large and well-conducted) studies have shown that antioxidants do not protect us from heart and age-related illnesses and indeed, why most anti-oxidants actually increase the average mortality rate. This was the conclusion of a major study that looked into 68 earlier studies in which more than 232,000 patients took part, confirming that the majority of antioxidants have no effect on the prevention of heart and vascular diseases, cancer, Alzheimer's, liver cirrhosis, etc. It also stated that the antioxidants beta-carotene, vitamin E and vitamin A even increase the mortality rate.[83] We will discuss this in more detail in Group 6 of the food hourglass when we cover good and bad food supplements.

If green tea contains such a fantastic amount of anti-oxidants as the many advertisements would have us believe, then these antioxidants would weaken our cells' defence mechanisms, which would in turn lead to fatal consequences. No, actually green tea is healthy. But not really because of its antioxidants, but because it is mildly toxic! We call these mildly toxic substances *flavonoids*. It is often claimed that flavonoids are good for us because they are antioxidants (claimed not only by the popular media but sometimes also by researchers who are obviously not quite at home with free radical chemistry). Flavonoids are good for us, but not because of their antioxidant activities, but rather because they are mildly toxic. These flavonoids stimulate the cells in our bodies to go into defence mode and our cells start producing those proteins that protect us against toxic substances. This means that in the long term, the cells are better protected against damage that is actually worse than that caused by the green tea itself.

Moreover, substances in green tea also influence the functions of proteins that play a role in the onset of cancer. Thus, these substances work not because of their 'magical' antioxidant properties, but because of their reaction with the specific proteins in our cells which regulate the growth cycle of cells.[84] This is interesting for the following reasons: when these proteins become unchecked, there can be uncontrolled growth which could cause cancer. The substance EGCG in green tea also activates the protein P53, which ensures that cells that start to mutate into cancer cells destroy themselves.[85] Green tea also suppresses the creation of TNF alpha, a substance that causes inflammation[86-87,] and micro-inflammation is an ideal breeding ground for cancer. Without actually mentioning the word antioxidant, I have listed various mechanisms that demonstrate that green tea can inhibit the onset of cancer.

This is all very well in theory, but what about in practice? Are there any studies that show that people who drink green tea are at less risk of developing cancer? Well yes, there are indeed many studies that demonstrate a link. For example women who drink ten cups of Japanese green tea a day (small cups compared to tea cups in the West) have on average developed cancer nine years later than those women who drink less than three cups of Japanese green tea per day.[88] One study in which 70,000 patients took part, showed that those people who regularly drank green tea had 57 percent less chance of developing bowel cancer.[89] According to a study in which 49,920 people took part, men who drank more than five cups of green tea per day had 48 percent less chance of developing prostate cancer than men who drank less than one cup per day.[90] Dr Fujiki, a researcher into green tea and cancer and the director of the Saitama Cancer Centre (Japan), says the following:

> 'Green tea delays cancer onset and it prevents second primary tumors, recurrence of tumors and metastasis in those with a history of cancer treatment. One million Japanese cancer victims live from five to 10 years after the cancer diagnosis, and they need cancer preventive agents without toxic effects. In this sense, green tea or green tea tablets are a practical, inexpensive preventive measure with accumulated scientific data.' (*Source: Vitasearch, The experts speak: cancer and green tea*)

The anti-tumour effects of green tea go further than cancer; it seems that genital warts can also be healed. I mention this because genital warts can be the precursors to tumours; the only difference between a wart and a cancerous growth is that the cells constituting a wart have not mutated sufficiently to become full-blown cancer. Those people whose immune systems cannot fight genital warts over the years run the risk of those warts developing into cancer.

Genital warts are a very common complaint and are caused by a virus, the *human papillomavirus* (HPV). Approximately 80 percent of humans will at some time be infected by this virus. HPV plants its DNA in the cells of the cervix or in the skin cells round the anus or the penis. Most people find that their immune systems will clear up these viruses but for some people the viruses will continue to proliferate (because they are so robust or because the immune system is not strong enough to combat this type of virus) and genital warts will develop, including some types which over time can develop into cancer. This is why the government advises women to have a cervical smear every two years, to check that there is no HPV present that could lead to cervical cancer.

What has all this to do with green tea, you may ask. The Food and Drug Administration (FDA) in the USA has recently approved a cream based on green tea that makes warts disappear.[91] This is remarkable because genital warts are hard if not impossible to cure. The most common method of 'removing' genital warts is to cut, burn or freeze away the infected cells. However, the warts will usually come back. Genital warts are a persistent problem for both doctors and patients. It is therefore very interesting that a simple cream based on green tea is able to remove these lesions. Remember that in the DNA of a genital wart, numerous mutations have already taken place that make the cells proliferate (otherwise they wouldn't be warts) so that with every step they get closer to being a cancer.

As well as reducing the risk of cancer, green tea also affects the metabolism. Green tea can help people lose weight and reduce the chance of developing diabetes, the pre-eminent metabolic disease.[92] Green tea improves blood sugar levels,[93] accelerates the burn-up of fat and improves the insulin sensitivity of the tissues. An increased sensitivity to insulin causes the liver, muscle and fat cells in our bodies to react faster to insulin and remove the sugar from the blood stream and store it. This means that sugar will cause less damage to other parts of the body.

Green tea also affects the brain. The amino acid *theanine* in green tea stimulates more *alpha waves* in our brains, so that we become calmer and can concentrate better. Green tea also reduces the coagulation of the blood platelets and that is good for the blood vessels in the heart and the brain. By drinking three cups of green tea a day you can reduce the danger of having a stroke (bleeding or a sudden blockage of an artery in the brain): a major study in which over 195,000 people took part showed that people who drank 3 cups of green tea per day had 21 percent less chance of suffering a stroke. When they drank 6 cups of green tea, then the risk of a stroke was 40 percent lower.[94]

However it's not all a bed of roses. Although the effect is small, green tea increases the risk of cancer of the bladder. This is due to the mildly toxic content of green tea as we mentioned earlier, which is concentrated in the bladder before it is expelled when we urinate. A low concentration of toxic materials can be healthy, but when the concentration is too high, it is not good for us and can cause cancer of the bladder. However the many advantages of green tea outweigh the slightly increased risk of bladder cancer.

Green tea, and tea and coffee in general, can also irritate the stomach lining. Drinking too much tea, especially on an empty stomach, can cause digestive problems for some of us. The best time to drink tea is one hour after a meal. Avoid drinking while you are eating because this will dilute the stomach acid and the digestive juices, which will hinder complete digestion and overburden the stomach and intestines.

We have now talked about two types of tea, namely green and white tea. There is another tea which research has shown to have beneficial effects on the body: ginger tea. An important constituent in ginger tea is *gingerol*. Gingerol curbs inflammation, which can lead to inflammatory illnesses such as rheumatism[95] and even cancer.[96–97] You can prepare ginger tea by steeping a few pieces of ginger in boiling water for about ten minutes.

*   *   *

So, green tea, white tea and ginger tea all have numerous beneficial effects on your health, with green tea in the forefront. My advice to those who want to lose weight or just generally keep healthy is to drink one to three cups of green tea per day, alternating with white or gember tea if you fancy it. For those people who drink green tea purely to lose weight, it is important to realise that they won't lose weight straight away just by drinking green tea. It will take many weeks before the metabolism will burn fats and sugars in a different way.

Soft drinks and shop-bought fruit juices

As we have mentioned time and again in this book, too much sugar is bad for the body. Too much sugar not only accelerates ageing but also increases the chance of cancer and, as more recent studies have shown, of vascular disease.

Food manufacturers have discovered a wonderful method of making us consume large amounts of sugar without us being aware of it. The result is known as, 'soft drinks', or even better, 'fruit juice'. The average can of soft drink contains 35 grams of sugar: the equivalent of 9 sugar cubes! Shop-bought fruit juices are not much better, despite them giving the impression of being healthy because they are made from fruit, or are labelled with enticing 'no preservatives' or 'added vitamin C!' slogans. But they are packed with sugar. The average shop-bought bottle of fruit juice contains 11 spoonfuls of sugar! Some manufacturers announce in capital letters on their juices 'No added sugars'. What they have in fact added is highly concentrated apple juice that consists almost entirely of sugar. However, because they have not added pure sugar (actually glucose) they can use the slogan 'No added sugars' in order to suggest that their product is healthy.

It isn't just the amount of concealed sugar in soft drinks and shop-bought fruit juice that is unhealthy. The package-ing of the sugar in these drinks doesn't do the metabolism much good either. Because the sugar is diluted in fluid, it is absorbed straight into the gut. This results in large sugar peaks in the blood, followed by insulin peaks. The insulin peaks exhaust the pancreas and make the cells of the body less sensitive to insulin, which can cause diabetes. The well-known Nurses' Health Study has shown that women who drink one or more soft drinks every day have 83 percent more risk of developing diabetes.[98]

As well as the large amounts of sugar, soft drinks such as Coca Cola also contain *phosphates*. Phosphates can cause accelerated ageing. Research carried out on mice has shown that mice with high concentrations of phosphates in their blood age more quickly and die earlier. Mice that have a diet containing a large amount of phosphates had a 25 percent reduction in their lifespans.[99] Phosphates are not only found in many soft drinks, but also in all kinds of industrially produced foodstuffs (for example, frozen pizzas, fast-food, cakes and biscuits).

Even the so-called 'healthy' light or diet soft drinks aren't that healthy. Diet coke, for instance, contains large amounts of phosphates. Moreover, diet soft drinks add extra weight to a person. People who drink diet soft drinks on a regular basis, have about twice as much chance of gaining weight. Another study demonstrated that the waist circumference of people who drank two or more diet soft drinks a day, increased five times quicker than that of people who didn't drink diet drinks.[100]

How is it that you still gain weight when drinking diet soft drinks that contain hardly any calories? A regular can of soda contains 140 kilocalories, while a diet can of soft drink only contains 1 kilocalorie (a person needs to ingest about 2100 kilocalories a day). However, diet soft drinks contain artificial sweeteners, such as aspartame. Unlike sugar, these artificial sweeteners contain very few calories, and consequently these diet soft drinks contain few calories as well. But these artificial sweeteners activate all sorts of neurologic and metabolic mechanisms in our body which cause people to gain weight, and even run a higher risk of diabetes. People who drink one or more diet soft drinks a day have a 67 percent higher chance of getting diabetes, according to a large study.[101]

With all these concealed sugars, phosphates and artificial sweeteners it won't surprise you to learn that soft drinks and other (mass-produced) fruit juices are one of the driving forces behind the obesity epidemic, particularly in the developing countries. The most overweight people no longer live in the USA, but in emerging, developing countries such as Egypt and Mexico. This is where huge amounts of soft drinks are consumed (because they are so cheap and taste good) and there has been an upsurge in obesity, heart and vascular disease and diabetes. In 1989, two percent of the population of Mexico had diabetes. In 2000 this figure had increased seven-fold to 15 percent and the trend is rising inexorably.[102] In Japan, where various cancers and heart and vascular diseases occur five to ten times less and where the population has the highest life expectancy in the world, we see that the amount of soft drinks consumed is ten times lower than that in the USA. The average American drinks 216 litres of soft drinks each year; Japanese drink no more than 22 litres.[103]

Of course, the fact that there are fewer occurrences of chronic illnesses such as diabetes, heart and vascular disease in Japan is not due just to the reduced consumption of soft drinks. However, consuming fewer soft drinks and shop-bought fruit juices is a vital part of a healthy eating habit. The diagram shows that drinking habits vary enormously from culture to culture. So there is a lot of room for improvement.

Soft drinks and shop-bought fruit juices do indeed belong on the top rung of the 'eat less' triangle. Replace these unhealthy drinks with water and green and white tea as much as you can.

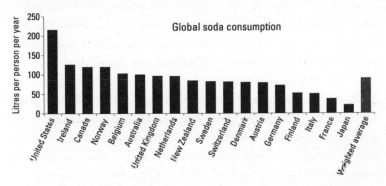

Source: *Global Market Information Database*, Euromonitor

Alcohol

The first thing that my professor of gastroenterology (the study of the digestive system) told us in his lecture about alcohol was that his hospital department would have two thirds fewer patients if alcohol had never been invented. Alcohol is a highly toxic material for the body that can damage the liver, heart and brain. Moderate alcohol consumption is not problematic, but excessive drinking brings countless dangers.

The best-known side effect of alcohol is liver cirrhosis. Cirrhosis means that the cells in the liver die and are replaced by tough connective tissue. Contrary to what most people think, cirrhosis is not caused so much by the direct effect of alcohol on the liver, but by another far more interesting mechanism.

Alcohol damages the intestinal wall and the stomach wall (those people with hangovers looking in the bathroom mirror are lucky they can't see the red, swollen lining of their stomachs). Because alcohol damages the lining of the gut, all kinds of toxic materials that have been produced by the bacteria in the intestines leak out of the gut into the bloodstream. Blood coming from the gut always first passes through the liver but the liver is well prepared. In the liver, between the liver cells, there are white blood cells that are equipped to remove toxic materials from the blood. However, because the intestinal walls have been damaged by alcohol, an excess of toxic materials passes directly from the gut into the bloodstream so the white blood cells in the liver become over-activated. The white blood cells think that there is a serious infection going on and produce inflammatory substances. These inflammatory substances can damage the surrounding liver cells and stimulate the surrounding connective tissue cells. These connective tissue cells start to produce connective tissue so that the liver becomes a mass of this tissue and thus turns 'cirrhotic', which is an irreversible process. As the liver is the body's biochemical factory, this is not a pleasant prospect.

But the liver and the gut are not the only organs damaged by alcohol. Alcohol can also damage the heart cells, which is why people who drink too much alcohol run the risk of heart failure. In addition, alcohol is bad for the pancreas; together with gallstones, alcohol is the primary cause of pancreatitis (an inflamed pancreas). Brain cells are not keen on alcohol either: in the long term, alcoholics can develop Wernicke and Korsakoff syndrome, two illnesses that are characterised by permanent memory loss, dulling of the senses and strange eye movements.

Of course, cirrhosis of the liver, an inflamed pancreas and alcoholic heart failure occur only in the case of long-term and chronic alcohol abuse. However, these illnesses demonstrate that too much alcohol is not healthy and can damage our bodies. Most of the damage can be repaired, but that doesn't detract from the fact that alcohol can quickly become a heavy burden on our bodies. Drinking five glasses of alcohol does not mean that your liver will fill with connective tissues or your heart will fail, but there will be damage at the molecular level, which in the long-term is bad for your health and can accelerate ageing. Most of the damage occurs through acetaldehyde, the degradation product of alcohol and a highly toxic material. It adheres to various proteins in the cells and to the cell membranes so the cells cannot function properly. Cells then become damaged and age more quickly.[104] Acetaldehyde is also the substance that causes hangovers, which make us feel sick, tired and listless after drinking too much alcohol.

Plus: alcohol will make you fat. One glass of beer contains about 105 kilocalories. Each of us needs an average of about 2,000 kilocalories a day (1,800 for a woman, 2,200 for a man). After drinking two glasses of beer you will have gained 210 kilocalories! One glass of wine contains about 85 kilocalories – of course, a glass of wine contains less liquid than a glass of beer. Each 100 ml of beer contains 43 kilocalories; 100 ml wine contains 85 kilocalories. Wine therefore contains twice the amount of calories as beer. Spirits are the front-runners when it comes to calories: whisky, vodka or gin have three times the amount of calories of wine.

What's more, alcoholic drinks don't make you fat just because they contain so many calories, but also because alcohol affects various metabolic mechanisms in the liver and the fatty tissue, which makes us pile on the pounds even faster.

The toxicity of alcohol can also be a good thing. We have already seen with green tea that mildly toxic substances can be beneficial, for example because they stimulate our cells' defence mechanisms. In fact, moderate alcohol use can be good for us. Research has shown that drinking one, or at most two, glasses of alcohol a day is beneficial for the heart and blood vessels. Men who drink one or two glasses of alcohol a day have 35 percent less chance of having a heart attack than those men who drink no alcohol at all.[98] Researchers were shocked by this large percentage. However, three or more glasses of alcohol pose a definite increased risk of high blood pressure, hardening of the arteries, heart attacks, inflammation of the pancreas, cancer and strokes. So, a moderate use of alcohol is permitted. Many health institutions recommend a maximum of two glasses of alcohol a day for men, and one glass for women. The rules for women are somewhat stricter because research has shown that already with two glasses of alcohol a day the risk of breast cancer is demonstrably higher. Women who drink two glasses of alcohol have 25 percent more chance of developing breast cancer. In short, alcohol has a very narrow 'therapeutic margin' for men and women alike.

One important point is that the alcohol intake must be spread out. So, don't stay off the alcohol for five days then have a celebratory Saturday night with five glasses. Try drinking one glass on three or maximum five days of the week. The Royal College of Physicians recommends that healthy people have 2–3 alcohol free days a week, so that the liver can rest. A good guideline can be to only drink alcohol during social gatherings, so that you'll only drink alcohol a few days a week. Be careful drinking alcohol at night: sometimes it makes you fall asleep faster, but your sleep is not as deep, causing you to be less rested the next day.

Diabetes patients or people who have trouble losing weight need to be careful with these recommendations: it is better for them not to consume too much alcohol. Alcohol puts a burden on the liver (the liver has to break down the alcohol) and people with diabetes already have an overloaded liver. In diabetes patients, there are already too many sugars and fats circulating in the blood that need to be broken down by the liver, causing a fatty liver. Add alcohol to it, and the liver becomes completely overloaded. People with diabetes should therefore drink little alcohol. And the same goes for people who want to lose weight. As we have seen, alcohol makes you gain weight. When a man drinks two glasses of wine after his meal every day, at the end of the year 72,000 kilocalories are added, which equals several kilos of extra weight. That's why people with diabetes or people who want to lose weight should not drink more than 2–4 glasses of alcohol per week. They should also consider going without alcohol for some time, to see what the consequences are for their sugar levels and weight.

As we shall see further on, there are better ways of ensuring your heart's health. It is, however, interesting to see that studies prove that it doesn't matter which type of alcohol you drink: wine, beer or spirits, they all have almost the same effect in reducing the risk of a heart attack. According to certain researchers, red wine would be slightly more beneficial to your health than other types of alcoholic beverages. The popular press is always claiming that red wine is good for you because it contains resveratrol. It's hard to find any article about health and wine that doesn't extol resveratrol as some kind of miracle substance that delays ageing. Indeed, some (but not all) studies claim that resveratrol increases the lifespan of yeast, worms, fruit flies and, according to one study, even one type of short-lived fish. The result of all this is that resveratrol has become one of the best-selling anti-ageing food supplements in the USA. Unfortunately, further research has shown that resveratrol does not increase the lifespan of mammals; mammals don't live a single day longer because of resveratrol.[105] However, resveratrol can increase the lifespan in mice that are given an unhealthy diet that contains large amounts of calories.

This has been demonstrated in a well-known study published in Nature. In the diagram you can clearly see how mice on a high-calorie diet die much earlier (bottom line):

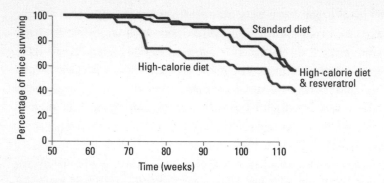

Resveratrol only works for mice if they have been put on an unhealthy calorie-rich diet. Resveratrol does not increase the lifespan of healthy mice. *Source: Resveratrol improves health and survival of mice on a high calorie diet*, Nature, 2006

the mice literally eat themselves to death. The mice given a normal diet live longer (top line). However, the mice that are given a high-calorie diet plus resveratrol live pretty much just as long as the mice on a normal diet (top line).[106]

Resveratrol therefore will not make you live longer or delay the ageing process (as many food supplement manufacturers would have us believe), but it could give you back the years you lost when you were living an unhealthy lifestyle. So the situation remains 'in balance' and the actual lifespan does not increase. All in all, resveratrol is a very interesting substance, especially for those patients with an unhealthy lifestyle or who suffer from metabolic diseases such as diabetes, high blood pressure or heart and vascular disease. There is, therefore, potential in this molecule derived from the peel of blue grapes. Pharmaceutical giant GlaxoSmith-Kline certainly agrees with this, and has acquired Sirtris, a small university company that is researching resveratrol and has some very interesting patents, which it bought for $720 million.

A small note: the amount of resveratrol given as supplements to mice is much higher than the amount of resveratrol found in a bottle of red wine. In order to get the same health benefits as the mice, you would have to drink between 35 and 100 bottles of red wine a day. This is not very advisable. Red wine is beneficial because it contains more substances other than just resveratrol.

And what is the situation with coffee?

In spite of the many reports to the contrary, coffee is actually a healthy drink (although it does have its disadvantages, especially in the amount you drink). Coffee is made from a vegetable product, which means it contains substances that can have a beneficial effect on the body. It is unfortunate that many people living in the West have such an unhealthy diet that coffee has become the main source of these healthy plant-based substances.

Research has shown that coffee has a protective effect against various ageing diseases including Alzheimer's disease,[107] Parkinson's disease[108] and diabetes.[109] This is partly due to the fact that coffee activates the NRF2 protein in the cells. This protein acts as a kind of switch that activates other proteins that protect the cells against free radical damage. An excess of free radicals plays an important role in diseases such as Alzheimer's, Parkinson's and diabetes. It has also been shown that coffee can protect against various types of cancer. Although coffee does provide some protection against the majority of cancers that have been researched, it can also be a risk factor for certain cancers such as stomach cancer, lung cancer and prostate cancer – at least, according to some older and less well-researched studies.[110]

Apart from these benefits, coffee can have its disadvantages. Coffee plays a somewhat ambiguous role in heart and vascular diseases: some studies claim that coffee can help prevent the onset of heart and vascular disease, but others state that once heart and vascular disease has occurred, coffee can increase the risk of heart attacks.[111] Additionally, we all know that coffee is a stimulant, and can lead to withdrawal symptoms, such as a lack of concentration, muscle fatigue, sleep disorders and so on. Too much coffee also damages the stomach and gut lining, which can lead to acid reflux or poor digestion. Coffee also increases the risk of osteoporosis.

Coffee has its advantages and disadvantages, although the advantages seem to outweigh the disadvantages, especially when it comes to the various illnesses associated with ageing such as diabetes, heart disease and Alzheimer's. Therefore coffee is at the bottom of the food hourglass's recommendations pyramid, but coffee should still be drunk in moderation. I therefore recommend drinking a moderate amount (no more than 3 cups a day).

Milk and yoghurt

Milk products are not as healthy as we used to think. That might come as a shock to many of us. In this book we have discussed how the scientists at Harvard put milk and yoghurt just below the forbidden top section in their new food pyramid, and even suggested that these dairy products could be totally replaced by calcium supplements and vitamin D. They had good reasons for this.

First of all we have to understand that evolution never intended humans to digest milk products. This is not very surprising. We have only been drinking animal milk for the last 10,000 years, even though humans have existed for about 180,000 years and our distant ancestors for millions of years before that. So our ancestors roamed the African savannah regions and the tropical rain forest for millions of years without their daily portion of cows' milk. Our bowels and bodies were never intended to digest milk products. About 10,000 years ago, nature made an attempt to make animal milk more digestable, when there was a mutation in our DNA that allowed some people to more easily break down a particular sugar contained in milk called lactose. Most Europeans acquired this mutation, so that they were able to digest milk more easily, but about 10 percent of Europeans don't have this mutation and become ill if they drink milk. Europeans (and most white Americans) are almost unique in the world in that they have this mutation associated with the ability to digest milk. Approximately 75 percent of Africans and almost 100 percent of Asians do not have this mutation and they are therefore unable to digest milk. Nevertheless, even for the 90 percent of Europeans who can digest it, milk is not advisable.

A study with 7,500 participants that appeared in Neurology, found that those who drank a lot of milk had a 2.3 greater chance of developing Parkinson's disease than those who drank no milk.[112] Whether or not the milk contained fat was irrelevant, as was whether or not it contained calcium. It was drinking milk itself that increased the risk of Parkinson's.

Milk and dairy products could also increase the risk of cancer. Women who ingested milk products more than once a day had 44 percent more chance of developing ovarian cancer than women who consumed milk products three times or less per month – according to the 88,000 test subjects of the Nurses' Health Study.[113] Men who drank two or more glasses of milk each day, had almost double the chance of prostate cancer metastasising than those who drank no milk. In thirteen independent studies, a relationship has been established between drinking milk and prostate cancer.[114, 98]

Scientists suspect that milk can increase the risk of getting these cancers because it contains all kinds of natural growth factors. These growth factors increase insulin and IGF in the body. And the more growth, the higher the risk of getting cancer. In a way, it is quite logical for cows' milk to contain growth factors, since it is intended to stimulate the growth of calves. But when grown humans start drinking it, the consequences may be less beneficial, especially in the long term.[115] In light of these insights, some very well-known professors in the US are now questioning the government recommendation to drink three (!) glasses of milk per day. According to the official guidelines, not doing so would be unhealthy and deprive you of calcium, which would be bad for your bones. However, at the same time, hundreds of millions of Asians never drink milk and it is not exactly that they are all walking about on crutches.

Indeed, milk is even non-beneficial in respect of those ailments which we thought it helped: milk doesn't prevent you from getting osteoporosis! Some studies even show that milk increases the chance of getting osteoporosis. Osteoporosis is the weakening of the bones and occurs mainly in women after the menopause, but men and pre-menopausal women can also be affected. This is very odd. Weren't we always told that milk helped to create strong bones?

Around the start of the 1990s, studies began to appear which raised questions about this milk-strengthens-the-bones myth. These studies examined the milk consumption in different countries. This is how scientists discovered that, in those countries that consumed only small amounts of milk products, fractures and osteoporosis occurred far less frequently. So, women in New Guinea consume almost no milk products; American women on the other hand, encouraged by the American food pyramid and the dairy industry, drink thirty times more cows' milk than the women in New Guinea. In spite of their huge consumption of milk, American women have 47 times more chance of suffering hip fractures compared to the women in New Guinea.[116-117] Other, later studies have confirmed that milk does not protect us from osteoporosis and in fact can increase the chance of contracting the disease. A study in which 78,000 women took part showed that women who drank two or more glasses of milk a day had a 45 percent higher chance of suffering hip fractures than those women who drank less than one glass of milk per week.[118]

Now how is it possible that we are still told that milk is good for our body, and especially for our bones? And that advertisements and government campaigns are continually extolling milk as a wonderfully healthy drink? While I am writing this, there is another campaign running on the television that advises you to drink milk if you still want to be able to dance the flamenco in your eighties. However, right at the end of this advert you might catch a fleeting message that the film is supported by subsidies from the milk industry and the European Union. The milk- and dairy industry is a multi-million dollar industry and that is why the government goes all out to convince us that milk products are healthy – obviously successfully. Harvard Professor Walter Willett, a world authority in the areas of food and health has the following to say,

'Consuming plenty of dairy products is being portrayed as a key way to prevent osteoporosis and broken bones. But not only does this fail to fit the bill as a proven prevention strategy, it doesn't even come close. [. . .] What's more, dairy products pose several proven and potential problems.'

But if we advise against milk products, where do we get our calcium? One healthy source is vegetables. Vegetables contain large amounts of calcium. 100 kilocalories of curly kale contains 455 mg of calcium, while 100 kilocalories of milk contains 194 mg of calcium. Each of us needs about 1,000 mg calcium per day. However, if you have a healthy diet then actually half of this would be enough because your body can better absorb calcium and there will be less acid in the blood (which takes calcium from the bones). Fruit contains much less calcium than vegetables, but even an orange contains at least 60 mg calcium.

And what does research show us? Eating fruit and vegetables strengthens the bones (a larger 'bone mineral density') and provides good protection against osteoporosis.[119] In short, for healthy bones, don't drink milk but eat plenty of vegetables.

The food hourglass advises against yoghurt as well as milk. Yoghurt is yet another dairy product that is praised to the skies, primarily by the industry because 'it regulates the passage through the gut', 'helps with constipation' or 'strengthens the immune system', illustrated by radiant, smiling ladies dressed in white who rub their stomachs in delight. Yoghurt does indeed help some of us who suffer from constipation and helps the passage through the intestines. This is because yoghurt actually upsets the gut. When we eat yoghurt, our bowels will rumble because they are struggling to digest these dairy products, which of course helps constipation or speeds up the passage through the bowel. It can even give you diarrhoea. Dr Hiromi Shinya, one of Japan's most famous doctors who even treats the Emperor of Japan, is a fervent opponent of yoghurt and milk. Dr Shinya is a celebrated gastroenterologist who invented the colonoscope, a cable with a camera attached with which the intestines can be examined via the anus. In this way he has seen thousands of bowels from the inside and been confronted with the damage that can be inflicted by milk and yoghurt. In one of his books, Dr Shinya has said that he 'has yet to meet anyone who eats yoghurt every day and still has a healthy bowel'. Yet switch on the television and hear that friendly voice on the advert telling you how yoghurt (with probiotics!) is a blessing for the bowel.

For those of you who simply can't survive without your

daily carton of yoghurt, just try oatmeal porridge instead or soya yoghurt or soya dessert. Soya yoghurt and soya dessert contain vegetable proteins and ingredients that are much easier for the intestines to digest, and they have other positive effects on our health, as we will see later.

SUMMARY

Green tea is not healthy so much because of 'anti-oxidants', but rather because it contains mildly toxic **flavonoids** so that green tea:
- reduces the risk of certain cancers;
- reduces the risk of strokes;
- keeps the skin healthy;
- encourages weight loss by speeding up the metabolism.

White tea can reduce the **formation of wrinkles**.

Ginger tea represses **inflammation** in the body.

Coffee can have more beneficial health effects for our bodies than the health problems it can cause, as long as it is drunk in moderation (maximum of 3 cups per day). Coffee reduces the chance of a variety of ageing illnesses such as:
- Alzheimer's;
- diabetes;
- Parkinson's;
- most cancers;
- heart and vascular disease.

Coffee increases the risk of:
- osteoporosis;
- withdrawal symptoms (headache, sleeping problems, problems with concentration);
- cardiac arrhythmia (with high intake);
- irritation of the mucous membranes of the stomach and gut.

Continued overleaf

Shop-bought fruit juices and **soft drinks** are unhealthy because they:
- contain large amounts of liquid sugar that causes high sugar spikes;
- contain phosphates that can accelerate ageing;
- contain artificial sweeteners that will make you gain weight.

Nature never intended humans to digest **milk and (drinking) yoghurt**, even if they are lactose tolerant. Dairy products can increase the chance of:
- Parkinson's;
- osteoporosis;
- prostate cancer and ovarian cancer;
- unhealthy bowels.

This goes for the milk we derive from **animals**, but not for mother's milk (through breast feeding) which has a totally different composition, compared to cows' milk. Breast feeding is beneficial to the child.

The following **alternatives** to milk contain plenty of calcium:
- vegetables (especially broccoli and vegetables with dark green leaves such as cabbage or spinach);
- tofu;
- soya dessert and soya yoghurt enriched with calcium;
- vegetable milk (soya milk, rice milk, almond milk, . . .) enriched with calcium;
- cheese (why cheese is an exception to dairy products will be discussed further on).

A few tips:
- drink enough, about 2 litres a day;
- drink 1 to 2 glasses of water 15 to 20 minutes before breakfast;
- drink mostly water, and further green tea as well as white or ginger tea;
- a maximum of 1(for women) to 2 (for men) glasses of alcohol a day, but let your liver rest every now and then. People who have

diabetes and people who want to lose weight should drink less;
- do not drink milk or (drinking) yoghurt.

Step 2: Vegetables, fruit, oatmeal, legumes and mushrooms

Bread, potatoes, pasta and rice

replace with

- oatmeal porridge (to replace bread), legumes (beans, peas, lentils, soya beans), mushrooms
- fruit
- vegetables

About oatmeal porridge and other starch substitutes
Probably the most revolutionary aspect of the food hourglass is that it advises against eating bread, potatoes, pasta and rice. In contrast to the usual way of thinking, the best way to lose weight is not by eating less fat or by eating lots of protein (as advised in low fat and high protein diets), but by simply cutting out or at least reducing your intake of bread, pasta, rice and potatoes. Even those who suffer from type-2 diabetes can 'cure' their 'chronic' diabetes by cutting out these starchy products in the sense that they will be able to inject less or even no insulin.[20–23] As a young medical student, I thought that was miraculous.

Of course, many people will have to take a deep breath when they hear that it would be better to cut out bread and potatoes. These products form a major part of the westerners' diet. But it really isn't difficult to take these starchy products off the menu. Take bread, for example. Most people in Europe and the US eat two cold bread meals and one hot meal a day. I would suggest changing to one cold meal a day: breakfast. The other two meals can be hot or cold and should mainly consist of various kinds of vegetables, peas, beans, tofu, mushrooms, Quorn or fatty fish.

However, we must still sort out that morning meal. You could replace bread with different kinds of fruit, nuts or dark chocolate, or with oatmeal porridge. Oatmeal porridge is made from oatmeal and is therefore a cereal product. Oatmeal is the only grain product recommended in the food hourglass.

Unfortunately, oatmeal is still relatively unknown and that is a shame because it isn't only good for us, but can also help us to lose a lot of weight. I could just as easily have called this book *The Great Oatmeal Diet*. Oatmeal contains a large amount of fibre that stimulates peristalsis in the small intestine. This is the rhythmic contraction of the small intestine that forces the food through the intestines. Fibre also prevents the sugars ingested during meals from being absorbed too quickly into the bloodstream, so a fibre-rich meal will ensure a lower blood glucose level.[120] For example, a study demonstrated that when diabetes patients eat more oatmeal porridge, they need to inject 40 percent less insulin.[121] The importance of fibre cannot be over-emphasised, because if there's one thing that all the studies that look at the connection between nutrition and health agree on, it's that healthy nutrition contains large amounts of fibre. Oatmeal also contains many elements that positively affect the blood vessels and the metabolism. The best known of these are the *avenanthramides* that slow down the hardening of the arteries and lower blood pressure.[122–124] The fibre, the avenanthramides and countless other ingredients in oatmeal make it a particularly healthy treat. What is perhaps even more important: because eating oatmeal will stop you eating bread, you will avoid insulin peaks and the daily dose of sugars that will slowly but surely pile on the weight.

Oatmeal porridge is usually prepared with milk, but because the food hourglass advises against this, you could use vegetable milk (soya milk, rice milk or almond milk). The soya milk sold in the supermarket often contains large amounts of sugar but there are other soya milks available that have no added sugar. Oatmeal porridge served with sugar-free soya milk may have less flavour, but you can solve this by adding a pinch of cinnamon or *stevia*, a healthy natural sweetener. However, even oatmeal porridge served with soya milk and some sugar will still help you to lose weight.

Oatmeal porridge is prepared by boiling the oatmeal in vegetable milk, like soya milk. Tip: make several portions of porridge so there will always be some available if you are hungry. You can keep the cooked porridge in the fridge and eat it hot or cold – you can easily warm it up in the microwave. For breakfast, you can have fruits, nuts or dark chocolate with your porridge. So, instead of starting the day with bread and ham or jam, you can start with a bowl of oatmeal porridge, together with:

- a bowl of strawberries, blueberries or raspberries;
  a pear or a banana;
- a handful of walnuts;
- a bunch of blue grapes;
- a piece of dark chocolate;
- a cup of green tea with lemon or a glass of freshly squeezed fruit juice.

This is breakfast as it should be. Another advantage of a regular, fruit-rich breakfast with oatmeal porridge is that when you look in your fridge in the morning, you don't have to think about what you are going to eat: there is your bowl of oatmeal porridge and some fruit for a delicious, healthy breakfast.

*   *   *

Here are a few other remarks about breakfast: don't eat breakfast cereals such as muesli or cornflakes. First of all, these are eaten with (cow's) milk, which we advise against. Moreover, breakfast cereals are sugar-bombs that cause high insulin peaks. They deliver anything but suficient amounts of vitamins, minerals and phytochemicals such as flavonoids. I also advise against those breakfast cereals that are supposed to be 'good for your figure'; they are full of sugar and are anything but a complete healthy breakfast. Just because the model on the box is pretty, slim and smiling doesn't mean that the cereal inside is good for you.

I once had a student who came to see me because he felt tired and faint every morning at around 10 o'clock. He had already talked to other doctors and was really worried. After I had asked him a few questions about his eating habits, I found out that in the mornings, somewhere around 8 o'clock, he would usually eat breakfast cereals. I advised him to stop this habit. Once he stopped there were no more attacks of fatigue or faintness. These attacks occurred because the breakfast cereals caused the glucose level in his blood to peak in the mornings. Thus he had two hours of energy, and then just as quickly as the sugar rush arose, the glucose level would fall at around 10 o'clock, which is why he felt tired and weak. This happens to many people every morning.

White bread also causes high glucose peaks. White bread is not actually a food. White bread is what remains when you have removed every mineral and every ounce of fibre and nutritional content from the bread. In the long term, white bread drastically increases the risk of diabetes and heart and vascular disease, and will even give you wrinkles![125]

Breakfast cereals and white bread are unhealthy because they contain so few nutrients and above all, they cause a peak in the glucose levels in your blood. These peaks are so important in nutritional science that scientists thought of a way to describe these sugar peaks. This description is called the *glycaemic index*. The glycaemic index is a measure of by how much a particular food will increase the glucose level in your blood. The higher its glycaemic index, the more unhealthy it is. The glycaemic index for glucose is 100; this is the reference point. Pure glucose goes straight into the intestine and from there is absorbed straight into the bloodstream, as it does not need to be cut into smaller pieces by the digestive enzymes. Bread, breakfast cereals, potatoes, rice and pasta all have a high glycaemic index. Wholemeal bread, brown rice and wholemeal pasta have a lower glycaemic index, although they are still relatively high. The glycaemic index of foodstuffs of 50 or over is considered to be high.

Foodstuffs with a high glycaemic index (above 50)	Glycaemic index
Glucose	100
Sugar (sucrose)	70
White bread	70
Soft white bread (hamburger buns)	85
Brown bread	65
White rice	70
Cooked potato	75
Brown rice	60
Chips	95
Mashed potato	90
'White' pasta	55
Wholemeal pasta	50
Cornflakes	85
Biscuits	70
Crisps	80

Foodstuffs with a low glycaemic index (less than 50)	Glycaemic index
Wholemeal rye bread	40
Orange, pear	35
Apple, peach	30
Dark chocolate	22
Cherry, plum, grapefruit	21
Fructose, soya beans	20
Tofu, walnuts, apricots	15
Lettuce, cabbage, broccoli, courgette, onion, garlic,tomato, aubergine	10

Notice that the glycaemic indices for cooked potato, mashed potato and chips are even higher than that for ordinary sugar. The sugars in these foodstuffs are separated out during the cooking process, so that they are absorbed into the gut very quickly, resulting in high glucose peaks. That's why researchers from Harvard University put potatoes in the forbidden top of their food pyramid, next to soda drinks and sweets. White bread, cornflakes and crisps also cause high sugar peaks.

If you want to eat healthily, eat as many foods as possible that result in the lowest glucose peaks in the blood. There are some comprehensive tables with lists of foodstuffs and their specific glycaemic index. Some of these tables also state the *glycaemic load*, as well as the glycaemic index . This *glycaemic load* is even better at predicting sugar peaks within the blood. Beetroot, for example, has a relatively high glycaemic index but a low glycaemic load, which means that its still healthy. However, for most of the food products it suffices to check the glycaemic index, if you want to see what kind of sugar spikes caused by this foodstuff. If you eat healthily and use the foodstuffs recommended in the food hourglass then you won't have to worry too much about all these tables, glycaemic indexes and loads. The food products in the food hourglass are primarily the ones with a low glycaemic index and load, so that any blood glucose peaks will look more like the rolling hills of Tuscany than the towering peaks of the Himalayas.

Those of you who want to lose weight or live a healthy life style like Professor Cynthia Kenyon (the worm lady), then I would advise you not to eat (or to eat as little as possible) bread, potatoes, pasta and rice. If you still want to eat bread then I advise you to try wholemeal bread (especially wholemeal rye bread), which has a lower glycaemic index, and the same goes for pasta and rice. There's a good reason for the researchers at Harvard to put 'white' pasta and white rice in the forbidden top portion of their food pyramid.

Some readers will have spotted that the glycaemic index for brown bread differs from that of wholemeal bread: the index for brown bread is 65, while that for wholemeal bread is 40. This is because brown bread is not the same as wholemeal bread. The difference between white, brown and wholemeal bread is a science in itself. If, however you are familiar with the composition of the humble grain, then you will know how many different kinds of bread come out of the bakers' ovens. Most bread is actually made from grains of wheat. Each grain consists of three parts: the endosperm, the bran and the germ.

The endosperm is the largest part of the grain. It is covered by a fibre-rich layer – the bran, (consisting mainly of fibre). The germ (containing lots of vitamins and minerals) is inside the endosperm (which is mostly starch). White bread is made purely from the endosperm; brown bread is made from the endosperm and a small part of the fibre-rich bran and the vitamin-rich germ. However, wholemeal bread uses the whole grain of wheat, in other words the endosperm, plus

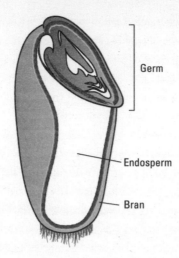

Composition of a grain of wheat.

the fibre-rich bran as well as the nutrient-filled germ. Brown bread can sometimes be simply white bread with a caramel-like food colourant added.

Conclusion: brown bread is not as good for us as wholemeal bread, because brown bread contains less nutrient-packed germ and fibre. So, because brown bread (and certainly white bread) contains less fibre, the glucose molecules that build up the starch in brown bread will be absorbed more quickly by the gut. This is why brown bread causes higher glucose peaks than wholemeal bread. That is to say, most of the time. Because sometimes the glycaemic index of wholemeal bread can be just as high as that of brown bread, or even of white bread. That is because some wholemeal breads are made of genetically manipulated wheat. This wheat contains more 'super starch', such as amylopectin A, that causes high sugar peaks. This type of bread is specifically made in the United States, and this causes even wholemeal bread to have a high glycaemic index. That is why American glycaemic index tables cannot always be compared to European tables, where wholemeal bread often is still genuine wholemeal bread.

Wholemeal bread notwithstanding, it's still better not to eat bread at all and replace it as much as you can with oatmeal porridge with strawberries, blueberries, apples, pears, dried fruit, nuts and the other alternatives that can be found in the food hourglass.

Some people wonder whether oatmeal will get a higher glycaemic index when you cook it, in order to prepare oatmeal porridge. Since to make oatmeal porridge, the oatmeal flakes are cooked in (vegetable) milk, which can affect their structure. But this is not the case: research has demonstrated that uncooked oatmeal flakes and cooked oatmeal porridge both cause about the same amount of low sugar and insulin peaks.[126-127] You see, the fibre in the oatmeal are water-soluble and form a kind of thick gel in the gut, that makes sure the sugars are absorbed less fast. It doesn't make any difference if these sugars come from oatmeal porridge or from regular oatmeal flakes.

However, make sure you buy the regular oatmeal flakes and not the 'instant' oatmeal that has been so preprocessed that it will cause higher sugar peaks.

By the way, the glycaemic index of oatmeal porridge is 50 which is lower than that of brown, white and most wholemeal wheat bread, which makes it an ideal bread substitute (the glycemic load of oatmeal porridge is 7, which is very low compared to the glycemic load of wholemeal wheat bread, which is 30). However, the glycaemic index and load are not the only reasons why oatmeal is recommended. The American Diabetes Association and lots of well-known food experts recommend oatmeal for many other reasons (we have already discussed these reasons, like the fact that oatmeal contains water soluble fibre like beta-glucans, avenanthramides for the heart and blood vessels, improves gut health, etc.). This is why oatmeal got a health claim from the European Union, while (wholemeal) bread (and pasta, rice and potatoes) haven't a single one.

All right then, that's it for the non-bread breakfast, but what about the other two meals of the day? How can we replace the potatoes, the rice or the pasta? In the food hourglass you can clearly see that one group from the top pyramid can be replaced by the corresponding group from the bottom pyramid. Thus, you can replace potatoes, rice and pasta with legumes (beans, peas, lentils), mushrooms and of course extra vegetables. We will come to these foods soon.

SUMMARY

Drastically reducing potatoes, bread, pasta and rice:
– will result in a significant weight loss;
– inhibits the onset of various diseases associated with ageing;
– can mean that type-2 diabetics no longer need to inject insulin;
– is *not* a ketogen diet since there are still sufficient sugars present (from fruit, legumes, starchy vegetables, oatmeal porridge et cetera).

Switch to a day with:
1 breakfast without bread or breakfast cereals;
– replace the bread with oatmeal porridge (served with vegetable milk);
– eat the oatmeal porridge with (dried) fruit (apple, strawberry, raspberry, banana, grapes, blueberries, peaches, raisins et cetera), nuts or dark chocolate.
2 (hot or cold) meals with poultry, fatty fish, tofu or Quorn and:
– one or more type of vegetable;
– legumes (beans, peas, lentils, etc.);
– mushrooms (oyster mushrooms, button mushrooms, portobello, shiitake, enokitake, etc.);
– eggs.

The **glycaemic index** is a way of measuring **glucose and insulin peaks**. Eat as many products with a **low** glycaemic index as you can.

Products with a **high** glycaemic index are primarily responsible for the enormous increase in obesity, diabetes, heart and vascular diseases and the acceleration of ageing.

Examples of such products are bread, rice, potatoes, mashed potatoes, breakfast cereals, French fries, biscuits, crisps, etc.

Oatmeal is more wholesome than other types of grain because it contains a lot of water-soluble fibre that:
- enhances the peristaltic movements of the gut;
- causes lower sugar and insulin peaks;
- is beneficial to the heart and blood vessels (which is an official health claim for oatmeal).

Legumes (beans, peas, lentils and soya)
Legumes constitute an important part of a healthy diet. For a start, legumes stabilise the glucose levels in the blood. They also contain only small amounts of *methionine*, which inhibits the production of protein. Methionine is an amino acid, but not just any amino acid. As you saw in Chapter 2 regarding proteins, these are built up from amino acids. The first amino acid from which every protein is built, is methionine. Methionine is therefore the 'starting block'-amino acid, that is the first to be incorporated in the amino acid chain that constitutes every protein. The building of proteins cannot start without methionine, which is why it is the most important building block for creating all of the proteins in our bodies. If there is insufficient methionine, then fewer proteins will be made. However, as we have seen, a decrease in the production of proteins in the body results in a longer lifespan for organisms. Producing more proteins will mean that we grow and age. Indeed, studies have shown that mice on a low methionine diet live longer.[128] So, anyone who eats food

containing only small amounts of methionine, such as legumes, will also produce fewer proteins, which can delay the ageing process. Legumes can be excellent alternatives to potatoes, pasta and rice. One study with 1,879 participants showed that substituting one portion of rice with legumes on a daily basis resulted in 35 percent less risk of the metabolic syndrome (this syndrome is characterised by high blood pressure, insulin resistance, unhealthy lipid blood levels, being overweight and having a beer belly). As well as beans, peas and lentils, we also have soya. Soya is a collective term for soya beans and all the product made from soya beans, such as tofu, soya milk or tempeh. Actually, soya milk is made by grinding soya beans in water, and tofu is curdled soya milk. Soya contains *vegetable* or *phyto-estrogens*, which scientists call *isoflavones*. Estrogens are hormones that are involved in many of the body's functions and hormone-sensitive cancers. Because soya contains hormone-like elements, it can have an effect on various hormone-sensitive cancers such as breast cancer. Many studies have shown that soya reduces the risk of breast cancer, sometimes by as much as a quarter.[129-130] Whether or not soya can help those women who already have breast cancer has yet to be seen. In any case, food supplements containing high doses of concentrated soya extract are not recommended to women who have breast cancer (nor are they generally recommended to healthy women). It is therefore better to use soya in its natural form.

Because of the femal-hormone-like substances in soya, some men are afraid of 'becoming more feminine' by eating lots of soya. Of course, this doesn't happen, since these vegetable hormone-like substances are far less powerful than the hormone produced by the body itself. Studies have demonstrated that soya products don't affect the density of testosterone, the production of sperm, or other hormonal functions in men.[131]

Soya can also have a positive effect on heart and blood

vessels. According to one study, people who ingest 30 grams of soya protein a day have a 20 percent less chance of developing cardiovascular diseases.[132]

Soya milk is a healthy alternative to ordinary (cow's) milk. Soya milk contains vegetable proteins that are much better for us than the animal proteins in cow's milk. Vegetable proteins are healthier than animal proteins because they contain less methionine and sulphurous amino acids. Even 'standard' yoghurt can be replaced by soya yoghurt or soya dessert. Some people are allergic to soya, but luckily this only applies to a small percentage.

A highly suitable alternative to meat is tofu, which is also made from soya and is chock-full of (vegetable) proteins. However, there has been research that has shown a link between eating tofu and dementia in older people in Hawaii.[133] But this increased risk of dementia could be caused by the way in which Hawaiians prepare their tofu. Large doses of aluminium are added to the tofu in the local production process, which could explain the increased risk of dementia. As is the case with many metals, a build-up of metals in the brain can cause dementia-like symptoms. Recently, another study has surfaced where older people in Indonesia who eat large amounts of tofu have been seen to have an increased chance of developing dementia.[134] The researchers do not rule out the possibility that this increase could be put down to a toxic preservative in the tofu. However, this starts to seem odd. Although there are studies that state for young people tofu can reduce the risk of dementia in later life I would recommend eating tofu in small quantities, especially in the case of older people (the over-65s). Small quantities means 4 portions per week for example. Further research should give us a definitive answer some day.

But there are other types of soya, apart from tofu, that is to say the *fermented* soya products. These are miso, natto and tempeh. 'Fermented' means that bacteria and fungi are added to the soya beans, which causes them to be a bit 'pre-digested' by these bacteria and fungi. This results in fermented soy, such as miso, natto and tempeh. Tofu and soya milk, on the other hand, aren't fermented. Some researchers believe that these fermented soya products are even healthier than the non-fermented soya. Natto, in particular, is an interesting product. It contains several substances that are beneficial to the heart and blood vessels, such as nattokinase and vitamin K2.[135] Nattokinase can break down specific protein strings that occur in the blood and make the blood coagulate better.[136] Nattokinase can also break down other proteins, such as the proteins in the brain that cause Alzheimer's disease.[137] It still needs to be researched whether natto actually reduce the chances of getting Alzheimer's in humans.

In Europe, fermented soya products are not eaten on a large scale yet, while they are very popular in countries like Japan, the country where the population has the highest life expectancy in the whole world. In Western countries these products can often be found in specialised health food stores.

SUMMARY

Legumes (peas, beans, lentils, soya):
- stabilise the glucose levels in the blood;
- contain only small amounts of methionine, which reduces protein production;
- contain vegetable proteins which are healthier than animal proteins.

Soya:
- can reduce the risk of hormone-sensitive cancers such as breast cancer;
- has a positive effect on heart and blood vessels.

Non-fermented soya products: tofu and soya milk.
Fermented soya products (are somewhat 'pre-digested' by bacteria and fungi): miso, natto, and tempeh.

Tofu (made from soya beans):
– can reduce the risk of dementia in later life for younger people;
– according to two studies can increase the risk of dementia in older people.

Natto:
– contains nattokinase which makes the blood coagulate less;
– contains vitamin K2 that is healthy for the heart and blood vessels.

Ordinary yoghurt can be replaced by soya dessert or soya yoghurt.
Animal milk can be replaced by vegetable soya milk.
Red meat can be replaced by soya products such as tofu, miso, natto and tempeh, among other things.

Mushrooms (and Quorn)

Mushrooms are amazing. They are neither plant nor animal, as they belong to a unique classification: *fungi*. Because they are so strange, biologically speaking, they have a special flavour, which means they can be used as a very suitable supplement in countless dishes. They can even be used as an alternative to the potato.

Let's look at how mushrooms can affect the body's health. To start with, mushrooms have a cancer-preventing effect. This is because mushrooms consist of long chains of strange polysaccharides, which stimulate the immune system. This causes the immune system to be on the alert, detecting and neutralising cancer cells more quickly. In Japan, hospital doctors use the extract of mushrooms as a supplementary treatment for cancer.[138–139] In Europe, mushrooms bear an official health claim which states that they can stimulate the immune system. Mushrooms are also used as a cancer-preventive. People who eat mushrooms every day have at least 50 percent less chance of developing stomach cancer according to one study.[140] Another study, which appeared in the *International Journal of Cancer*, and in which over 1,000 women took part, showed that those women who ate 10 grams of mushrooms per day had 64 percent less chance of developing breast cancer, compared to women who ate no mushrooms. If these women also drank green tea every day, than their risk of breast cancer would reduce by 89 percent.[141] This is an excellent illustration of the synergy you get when you combine different lifestyles: mushrooms are good for us, green tea is good for us; the combination of the two is even better for us.

There are different types of mushrooms, so that there is something to suit every taste: from button mushrooms and oyster mushrooms to shiitake, portobello and enokitake. Oyster mushrooms are particularly well-known for their powerful anti-cancer effects and they can be prepared in so many delicious ways.[142]

Quorn is a food quite similar to mushrooms; it is made from proteins that come from a fungus. Because it contains lots of proteins, vegetarians use it as a substitute for meat.

In the food pyramid, we can see that both Quorn and soya can serve as meat substitutes because they contain so many proteins. Legumes and mushrooms also contain quite large amounts of proteins, but they also have plenty of carbohydrates, which means they can be the perfect replacements for potatoes, rice or pasta. In short, you can use Quorn and soya as well as legumes and mushrooms in place of potatoes or pasta, because all these products can bring more variety to (hot) meals. And of course, legumes, soya, mushrooms and Quorn don't cause such high glucose and insulin peaks as the carbohydrate-rich pasta or potatoes. It's those glucose and insulin peaks that are so bad for our health.

We often forget that potatoes, pasta and rice can also be replaced by a second portion of vegetables. After all, vegetables are the basis of our nutrition and of the food hourglass.

SUMMARY

Mushrooms stimulate the immune system so that they can provide a **defence against cancer**.

In Japan, doctors use mushrooms as a supplementary treatment in chemotherapy.

Quorn is made from a fungus and contains large amounts of proteins, which means it can be used as a meat substitute.

Protein-rich products that can be used as meat substitutes are:
– Quorn;
– soy (tofu, miso, natto, tempeh).

As possible meat substitutes, you will find soya and Quorn in step 3 of the food hourglass together with fatty fish and poultry. However these food products can also be used as substitutes for potatoes, pasta and rice.

Vegetables

In themselves, vegetables are not healthy: if there's one thing that vegetables hate, it's being eaten. So, vegetables are stuffed with all kinds of toxins, to protect themselves against greedy mammals, insects and birds. Most vegetables you find in shops won't contain enough toxins to make us ill, but they still contain toxins, and it's these very toxins that are so good for us. After all, they activate the defense mechanisms of our cells, so they can better resist the more damaging toxins that are created as a by-product of our metabolism. Vegetables form the base of the food hourglass. This is in contrast to the food pyramid and the food disc where the bases are bread and other starchy products. That's not good because our bodies have gone through millions of years of evolution during which, most of the time there were no such things as bread, potatoes, pasta and rice. Our bodies are however not made to eat large daily loads of bread and other starchy foods. The reason why we eat such a great deal of these starchy foods is that the food pyramid and the food plate were thought up in countries where the grain industry forms a large part of the economy. Producing grains is cheap; they can be produced in enormous quantities, and they are easy to prepare. The food industry loves grains! So should you too apparently.

Vegetables form the basis of the food hourglass because, from pre-history, people have eaten lots of vegetables and many substances in vegetables are good for our bodies. This is why studies have shown that various kinds of vegetables are effective against specific cancers, such as bowel cancer, the very prevalent prostate cancer in men, or cancer of the bladder. The anti-cancer effect is more pronounced with vegetables than with fruit (which is particularly good for heart and vascular disease, more on that later).

Certain elements in vegetables can inhibit damage to DNA and even reverse it.[143] Mutations in DNA cause cancer and ageing. For example, it seems that men who eat vegetables three or more times a week have 41 percent less chance of developing prostate cancer than those men who eat vegetables less than once a week.[144] Women who eat at least 1 kilo of broccoli per month, have 40 percent less chance of developing breast cancer than those who eat less than 350 grams of broccoli per month.[145] Women who eat tomatoes at least three times a week have 70 percent less chance of developing ovarian cancer.[38] Such studies are important because they hint at the power of food, and in this case vegetables.

٭   ٭   ٭

Vegetables are especially good in inhibiting age-related illnesses such as macular degeneration, cataracts, reduced memory function and cardiovascular disease. Macular degeneration occurs when the cells in the retina begin to die. This disease is caused by the build-up of waste in the cells of the retina. This build-up of waste in cells throughout our bodies is one of the reasons we grow old and die. Over time so much cellular waste accumulates that our cells choke and die. When that happens in the cells of the eye it is called macular degeneration. When this build-up takes place in our muscle cells, and they give up one after the other, doctors call it *sarcopenia*, the loss of muscle mass. This process occurs in older people and causes arms and legs to get thinner. When our nerve cells start to die then our sight and hearing become impaired and when this happens in the brain doctors call it dementia. Eventually, so many cells die that the whole organism will give up completely. That person will die of 'old age'.

Let's go back to macular degeneration which, together with diabetes, is the primary cause of blindness in the western world. At least 20 percent of the over-60s suffer from this disease. This number is increasing, because macular degeneration is a disease associated with ageing: one out of three people over the age of 74 suffers from macular degeneration. If we all grew old enough, everyone would eventually suffer this problem and nothing much can be done about it. However, people who eat vegetables at least five times a week halve their chances of developing macular degeneration.[4] Vegetables contain all kinds of substances that can prevent waste materials building up in the retina, so that the risk of developing this disease is cut drastically.

Heart and vascular disease are also illnesses of old age. Fruit has a prominent effect on the diseases of the heart vessels, but vegetables also play an important role. The following diagram appeared in *Nature*, and it shows that people who eat a great deal of green-leaved vegetables have 32 percent less chance of suffering a heart attack.

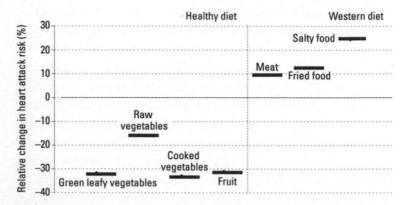

Green-leaved vegetables and broccoli reduce the risk of heart and vascular diseases by over 30 percent. *Source: Health: Edible advice*, Nature, *2010 (based on the Interheart Study)*

It's especially broccoli and the green-leaved vegetables such as salad, spinach and cabbage that have a pronounced effect on our health. Broccoli stands out and broccoli shoots in particular. Broccoli shoots are the young plants and contain high concentrations of all kinds of healthy substances, sometimes a hundred times more than the adult broccoli.

Now I can hear what some of you are saying: reduced risk of certain cancers if I drink green tea and eat mushrooms, less chance of a heart attack if I eat plenty of vegetables, less chance of macular degeneration if I eat broccoli or salad every day. What will actually kill me if I have such a healthy diet?

That is indeed a very interesting question. Those of us who have a very healthy diet would, in principle, mainly die of *intrinsic* old age. That is the ageing process programmed genetically in our genes. When you die from intrinsic old age then you will die at a very old age, for example, when you are eighty or ninety and after a short period of being 'chronically' sick. This period will only last for a few years and is typical for people who have lived a healthy life or have good genes. These people have a good life right up to the end and then things degenerate very quickly.

This is so different from the fate of people who eat lots of sugar, snacks, meat and fast food and who eat very few vegetables and fruit. These people will die primarily from *extrinsic* ageing. This sort of ageing is built up from external factors, mostly due to an unhealthy diet. These people tend to die at an earlier age, and usually after a long period of sickness and deterioration. This is how it usually goes: around the age of fifty, these people will start to suffer from high blood pressure, chest pains or shortness of breath due to the clogging up of the heart vessels, (pre-)diabetes, problems with the eyes, pain in the muscles and nerves, fatigue and loss of concentration. After fifteen to twenty years of reduced quality of life, many visits to the doctor and admissions to hospital, these people will die at around the age of 70, at home in their armchair or in a hospital bed.

In the diabetology department, I have seen many patients passing away like this. The scenario is usually as follows, patients have been lying in the hospital bed for weeks, looking rather dazed and apathetic due to all the toxic waste products which are building up in their blood and because their saccharified kidneys are having difficulty doing their filtering job. They are too tired to get out of bed because their coronary arteries are almost fully blocked despite the four stents that have already been placed. Their skin is pale with a yellowish hue and wrinkled because of all the sugar-crosslinking of the collagen in the skin. They have high blood pressure, heart problems, kidney problems, liver problems, stomach and bowel problems, problems with their joints and muscles and difficulty concentrating. Then the curtain falls and a calcified plaque breaks open in a large coronary artery of the heart, to which clotting proteins immediately attach themselves, so that the artery is suddenly completely blocked and the patient dies, usually without dilated pupils or any clutching of the chest, because the nerves which go from the heart to the brain are so damaged by the sugar that they no longer feel any pain when the heart finally gives out.

This is one way of dying that will happen more and more often if we continue to underestimate the importance of nutrition and preventive medicine in general, and if health professionals are not more trained in preventive medicine and nutrition. Naturally, we must not forget that some patients are in the metabolic diseases department because of a genetic susceptibility to diabetes and heart disease. In spite of this, the statistics don't lie. The number of people with metabolic illnesses has increased by several times in recent decades and currently 40 percent of Europeans will eventually contract type-2 diabetes at some point in their lives,[146] a disease that increases the risk of heart and vascular disease by a factor of five. The well-known Harvard professor Frank Hu wrote in an article in *The New England Journal of Medicine* that 91 percent of cases of type 2 diabetes could be prevented. Taking this together with the insight that a bad diet plays the biggest role in the onset of diabetes really provides food for thought.

Now we come back to the question that we just asked and that has still not really been answered: imagine you live a very healthy life so that you drastically decrease the risk of getting diabetes and heart and vascular disease, and imagine you are lucky enough (whether or not because of your genetic make-up) to grow very old; what will you actually die of? Those people who live to 100 or more (*centenarians*) do not die so much from heart attacks or cancer. They will die primarily from amyloidosis. Amyloidosis is the accumulation of proteins in the blood vessels and organs, a process that takes place throughout our lives (and speeds up if you ingest a lot of protein). This makes the blood vessels and organs fragile and eventually a blood vessel will burst, or a vital organ will fail, and many centenarians will die from a genuine disease of ageing even the most resilient elderly can't outrun.

Now, back to the vegetables. We have seen that vegetables can slow down many of the diseases associated with old age. But in what amounts? In the best-case scenario, we should eat 300 grams of vegetables a day, ideally half of them raw and half steamed or boiled. That seems like a lot, doesn't it? But the carbohydrate-rich foods such as potatoes and pasta will come to our aid. Now, because we must leave these products alone as much as we can, we can replace them with different vegetables, such as lettuce, aubergine, broccoli, cauliflower, cabbage, curly kale, sprouts, beetroot, tomatoes, courgettes, pumpkin, sweet peppers, chicory, radish, carrot and so on. Lettuce can be an ideal basic vegetable. Now I know that lettuce isn't exactly valued for its rich taste, but you can do something about that. It's worth trying, because lettuce can be made more appetising with ordinary vinegar, balsamic vinegar, tomato vinegar, raspberry vinegar, rice vinegar, wine vinegar and other countless, homemade vinaigrettes based on olive oil, linseed oil or walnut oil. You can also add so many healthy foods to a salad to improve the taste, such as onions, shallots, onion powder, orange juice, tomatoes, walnuts, linseed, small pieces of anchovy, herbs, thinly sliced apples or mandarins, peas, et cetera. In the supermarket you can find herb mixes to flavour chicken soup, but you can also use them to elevate salad to new culinary heights.

There are thousands of recipes that help to add flavour

to salad and other vegetables. At the end of this book you will find several recipes, such as broccoli dressing, made with broccoli, rice vinegar, a few spoons of mustard and a couple of cloves of garlic. Or how about a the mustard-garlic vinaigrette, made with balsamic vinegar, a dash of olive oil, a couple of cloves of garlic, pepper and a few herbs? Delicious!

I would advise you to go even further, and elevate the eating of vegetables to even greater heights, health-wise. You can do this by making a vegetable juice every day. One staunch advocate of a daily glass of vegetable juice is the 87-year-old American billionaire David Murdock. Murdock does everything he can to live a long (and healthy) life, so he eats lots of fruit and vegetables. He therefore swears by vegetable and fruit juices. His mother died of cancer at the age of 42, so he was (and still is) especially motivated. Murdock invested 500 million dollars in a research centre that studies the health-promoting ingredients in fruit and vegetables, and he introduced *salad days*, a country-wide initiative to remind American teenagers at *high schools* that there exists such something called 'vegetables'. It goes without saying that Murdock does not eat red meat or milk products and his menu consists mostly of fatty fish, beans, nuts and lots of fruit and vegetables and vegetable juices. In this way he intends to 'reach the age of 125'.[147]

The most famous vegetable juice is the concoction given to mice without immune systems by Dr Beliveau. Beliveau is a famous researcher into the substances contained in vegetables, spices and mushrooms that have an inhibiting effect on cancer. Dr Beliveau's mice were genetically manipulated to have no immune systems so they were highly susceptible to cancer because it's the immune system that clears away budding cancer cells. Dr Beliveau then injected the mice with cancer cells. The mice that were actually given Dr Beliveau's vegetable cocktail developed cancer much later, and the cancer grew more slowly compared to the other mice that developed huge tumours (in human terms, the tumours would have weighed several kilos). This is what is in Dr Beliveau's famous cocktail:

100g broccoli
100g sprouts
100g spinach
100g beetroot
100g shallots
100g garlic
100g cranberries
100g kidney beans
100g grapefruit
2 teaspoons turmeric mixed with 10 ml linseed oil
2.4g green tea extract (= 6 cups of tea)
2 teaspoons black pepper

It is very easy to make vegetable juices. Google 'vegetable juice' and you will find thousands of websites. Type 'vegetable juice' on YouTube and you can watch hundreds of films showing you how to make all kinds of vegetable juices. First of all, make sure you wash the vegetables thoroughly to get rid of any bacteria and pesticides. To make a delicious vegetable juice, you can try using: two carrots, half a stalk of fennel, three sticks of celery, half a cucumber, half a courgette, one beetroot and a few florets of broccoli. Put them all in a juicer and before you know it you will have a massive dose of flavonoids, vitamins and minerals ready to drink.

SUMMARY

Vegetables:
- form the basis of every healthy diet (as opposed to starchy products such as bread, potatoes, rice and pasta);
- reduce the risk of various cancers;

Continued overleaf

- reduce the risk of age-related illnesses (such as macular degeneration);
- cruciferous green vegetables, like broccoli, kale, Brussels sprouts, cabbage and arugula (rocket), reduce the risk of heart and vascular disease.

Vegetable juice:
- can considerably repress the growth of tumours in mice;
- is an ideal way to ingest extra vegetables.

Fruit

Fruit is good for us but vegetables are even better. It is more important to eat vegetables than it is fruit. This is why vegetables form the basis of the food hourglass and fruit is placed in the group above vegetables. But how healthy is fruit? From a study that lasted for 6.5 years and involved about 40,000 participants, it was shown that those people who were in the top quartile of people who consumed fruit (a quartile represents a quarter or 25 percent of the test subjects), 21 percent had less chance of dying than the people who were in the lowest quartile. Thus the researchers were comparing the 25 percent of people who ate the most fruit with the 25 percent who ate the least fruit. [148] Countless studies have shown that people who eat sufficient fruit are less likely to become ill, and live longer.

Fruit is good for us because, among other things, it slows down ageing. If you eat plenty of fruit you will (literally) stay looking younger: certain ingredients in fruit will slow down the formation of wrinkles and will inhibit the 'photo-ageing' of the skin. Photo-ageing is the ageing of the skin caused by sunlight. One well-known component in fruit in respect of the ageing of skin is ellagic acid.

Raspberries, strawberries, bilberries and pomegranates in particular contain ellagic acid, which is an acid that keeps the skin young by slowing down the degeneration of the

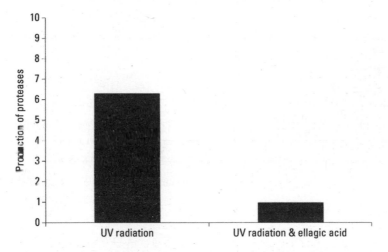

Ellagic acid

connective tissue, which keeps the skin more supple and elastic. Ellagic acid also slows down the ageing of skin by protecting the DNA in the skin cells against sunlight and by inhibiting the proteases.[149]

Sunlight (UV-radiation) activates the production of proteases, proteins that break down the collagen and elastin in the skin so that it becomes less supple and wrinkles form. A sufficient dose of ellagic acid drastically reduces the production of proteases.
Source: Dietary compound ellagic acid alleviates skin wrinkle and inflammation induced by uv-b irradiation, Experimental Dermatology, 2010

As we described briefly in the previous chapter, proteases are proteins that break down the collagen and elastin in the skin (so that the skin becomes less supple and more wrinkly). One thing that activates proteases is sunlight. In the diagram you can see that proteases are five times less activated when the skin is exposed to the sun than the skin cells that have been treated with ellagic acid.

Fruit is not only good for the skin, but also good for the blood vessels. It's no coincidence that fruit has a healing effect on the skin as well as the blood vessels. Just as the eyes are the window to the soul, the skin is the window to the blood vessels. Having healthy skin usually means that the blood vessels are also healthy. This is because skin is composed of the same elements as blood vessels: collagen and elastin. One way of looking at it is this: if we are old and wrinkly on the outside, we'll also be old and wrinkly on the inside because the collagen and the elastin deteriorate both in the skin and the blood vessels.

This connection between the condition of the blood vessels and the skin can also explain Frank's sign. Frank's sign is a small indentation in the earlobe. According to various studies, people who have this sign are five times more likely to have a heart attack than those who don't have it.[150]

Scientists still don't know why people with Frank's sign are more susceptible to heart and vascular disease. Some have put forward the following explanation: Frank's sign is a kind of wrinkle that occurs in the skin because the skin of a person who has Frank's sign is generally less healthy and will therefore wrinkle much earlier. The same goes for the blood vessels because they consist of the same components. The blood vessels, like the skin, are of poor quality, which means they will be damaged much more quickly. Clots can form on the damaged parts, which can lead to the hardening of the arteries (which makes these people susceptible to heart

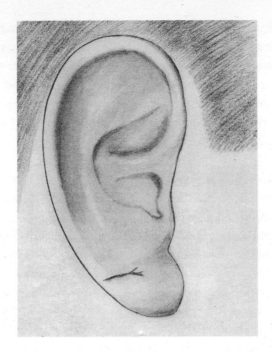

Frank's sign.

attacks). Briefly, the poor condition of the blood vessels is mirrored in the poor condition of the skin, but it is important to state that not all of those who have Frank's sign are doomed to have a heart attack.

Whatever the case, our skin is the mirror of our blood vessels, and this relationship explains why fruit is good for the skin as well as reducing the risk of heart and vascular disease. The same applies to fruit extracts in the form of freshly squeezed fruit juices. They will keep the blood vessels young. As we get older, the walls of our blood vessels become less supple and thicker. Test subjects who drank a glass of pomegranate juice every day for a year reduced the thickness of the intima and media (the two innermost layers of the walls of the blood vessel) by 30 percent, while the intima and media of the test subjects who drank no pomegranate juice got several percent thicker. The pomegranate juice also caused the oxidised LDL – the bad cholesterol particles which stick to the walls of the blood vessels and which cause the hardening of the arteries – to be reduced by 90 percent. Blood pressure also dropped by 12 percent, which by medical standards is remarkable.[151]

Fruit can also protect us from other diseases associated with ageing such as wrinkles and heart and vascular diseases. Fruit also has an effect on *the* ageing disease of the western world, which is Alzheimer's disease.

Alzheimer's is the result of the ageing of brain tissue. Over the decades, the brain cells become suffocated by coagulating proteins, which eventually damage the brain cells so much that they die. This happens sooner for some people than others. When older people (for example if they are in their 70s) suffer from early Alzheimer's disease, doctors will describe it as Alzheimer's. However, the coagulation of proteins affects us all. But the usual rule applies here as for all ageing diseases: if you wait long enough, we will all suffer from Alzheimer's. The statistics bear this out. From the age of 65, the risk of Alzheimer's doubles every five years so that eventually half of the people aged 85 or over will have Alzheimer's.

Researchers have tested countless substances to see if any of them can halt the progression from mild cognitive impairment to Alzheimer's. These are principally antioxidants such as vitamin C, vitamin E or beta-carotene. Studies have shown that these antioxidants have no effect on Alzheimer's disease. Even the much-lauded vitamin E, which in earlier studies seemed to be having an effect when used in high doses, turned out in the end not to prevent or hinder the progression from mild cognitive impairment to Alzheimer's disease.[152]

Researchers therefore decided to approach Alzheimer's differently and leave antioxidants for what they are. Instead they gave the test subjects fruit juices, because they contain large amounts of flavonoids. Flavonoids are partly responsible for the bright colours of fruit (the red of a strawberry, the blue of a blueberry, the orange of an orange). Flavonoids are much better for us than those antioxidants you can buy in the supermarket. It is often claimed that flavonoids are antioxidants, but flavonoids are actually good for us because they work in an entirely different way. The body often considers these materials to be mildly toxic and the body's own antioxidants systems are activated. These are far better able to combat free radicals than the antioxidants ingested in tablet form and which in fact, weaken the body's own antioxidant systems.

It has been shown that people who drink at least 3 fruit juices a week (or vegetable juice: vegetables also contain a lot of flavonoids) had 76 percent less chance of contracting Alzheimer's, compared to those who drank fruit juice less than once a week. People who only drank fruit juice twice a week had only 16 percent less chance of Alzheimer's. This clearly demonstrates that a particular health intervention only reaches its maximum potential when the intervention is applied on a regular basis. It was concluded that:

Fruit and vegetable juices may play an important role in delaying the onset of Alzheimer's disease, particularly among those who are at high risk for the disease. These results may lead to a new avenue of inquiry in the prevention of Alzheimer's disease. (*Fruit and vegetable juices and Alzheimer's disease*, The American Journal of Medicine, 2006)

Some people will, without doubt, be shaking their heads when they hear that people who drink fruit juices have 76 percent less risk of Alzheimer's and claim that this must have been a minor and poorly conducted study. Yet this study appeared in the *The American Journal of Medicine*, a major medical journal where the study design was carefully checked. Besides, this study was carried out using more than 1,800 test subjects and lasted for ten years. Of course we need to interpret this study carefully because people who regularly squeeze fresh fruit juices often also smoke less, exercise more, eat more vegetables, etcetera. Despite that the researchers accounted for these 'confounding factors', it's impossible to account for all potential confounding factors. Nonetheless, the message still remains the same: whether it's the fruit juice or other lifestyle fators: the risk of dementia can be considerably decreased by a healthy lifestyle. What's more, countless other studies about the consumption of fruit and fruit and vegetable juices have shown that they can reduce the risk of dementia, while at the same time this certainly doesn't apply to the use of vitamins and anti-oxidants via dietary supplements.[153] We will come back to this later on.

As we are discussing the positive effect of fruit on the brain, we absolutely have to mention blue and red fruits in particular: strawberries, raspberries, cranberries, blueberries, bilberries ('myrtle berries') and blackberries. They contain anthocyanins, a specific form of flavonoid that has an especially protective effect on nerve cells. Anthocyanins are natural colouring agents that give all these berries their red and blue colouring (anthocyanins are actually also partially responsible for the red-orange colour of autumn leaves).

In particular, blueberries and bilberries protect against the deterioration of nerve tissue.[154] Nerve cells are particularly sensitive to ageing because they are so metabolically active (brain cells use ten times more energy than other cells in the body) and because they have to last us all our lives (nerve cells hardly ever divide after birth). Old rats, which were put on a diet with extracts from blueberries for two months, functioned better in all the tests concerned with memory, balance and movement coordination. According to this study

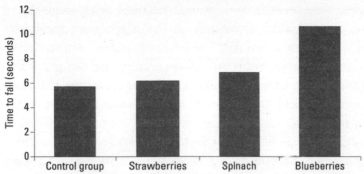
Walking along a rod

Old rats that are fed blueberries can walk further round an obstacle course than rats that are not fed any blueberries at all. These rats have better balance and coordination, properties which require a great deal of brain activity and which decline considerably as we get older. *Source: Reversals of age-related declines in neuronal signal transduction, cognitive, and motor behavioural deficits with blueberry, spinach, or strawberry dietary supplementation*, The Journal of Neuroscience, 1999

that appeared in *The Journal of Neuroscience* the rats that had been given the blueberries to eat had more energy and performed better in the cognitive tests.

Interestingly, British pilots who flew night bombing missions over Nazi Germany were given bilberry jam to eat. The deep blue colour of bilberries is due to the anthocyanins which can protect nerve cells and eye cells. Bilberries look very similar to blueberries. You can only see the difference between the two berries when you cut them in half: blueberries are green inside, bilberries are purple inside. This means that bilberries have more of the purple anthocyanins, and indeed research has proved that bilberries are even better for us than blueberries.

According to some scientists, even astronauts could benefit from drinking a blueberry cocktail, which could protect space travellers from high-energy radiation in space. (Astronauts are continually exposed in space to high doses of cosmic radiation because they are no longer protected by the earth's magnetic field.)[136] The major problem with a trip to Mars is that it would be very difficult to protect space travellers from the damageing cosmic radiation during a journey that would take years. Research has shown that blueberries can reduce the damage that radiation causes to cells. Rats that are fed with blueberries for two months before being exposed to high doses of radiation are better able to resist the radiation than rats on a normal diet.[155]

As well as these benefits, anthocyanins also have a cancer prevention effect. Extracts from berries suppress the growth and proliferation of eight different human cancer cell lines.[156] In another study, rats were given a substance that is highly carcinogenic; the rats that were given a diet which was 5 percent composed of one particular type of blackberry developed 54 percent fewer tumours than the rats which didn't have blackberries on their menu.

The skin and the sun
In this section about fruit, we have talked a great deal about the skin, and how substances in fruit will keep the skin healthy and slow down skin ageing. But this is just half of the story. One major cause of (premature) ageing of the skin is the sun. The effect of sunlight on skin ageing is of major importance, so let's discuss it.

The damaging effect of sunlight on the skin can be seen by doctors whenever they examine a patient, especially if that patient is older. For example, take the average 75-year-old woman. This lady has a considerably wrinkled face, and her hands are covered in liver spots, creasing and brown discolourations. However, during the medical examination if you look at the skin that is not exposed to sunlight, like the buttocks and bottom, you will see a large difference. This

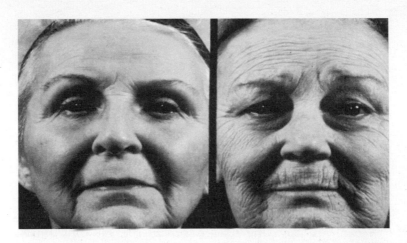

Two women aged 71. The lady on the left has sunbathed much less frequently than the lady on the right. *With the permission of Professor C.E.M. Griffiths, University of Manchester*

skin that is 75 years old has hardly seen any sunlight at all, is completely white, smooth and almost wrinkle-free. Obviously this skin doesn't entirely escape the ravages of time: it has lost some of its elasticity and is thinner because it has less collagen and elastin but even so, this skin is almost wrinkle-free and there are many fewer pigmentations such as moles or liver spots, certainly when compared to the skin of the face or the hands. When you compare these two skin areas you can clearly see the enormous effect of sunlight on the skin.

You can see the opposite effect in a dermatology department. When I was doing my internship as a medical student I occasionally came across former sun-worshippers, people who always love to look tanned and spend their entire holiday lying in the sun or hurry outside to catch every ray of sunlight. As a a 60- or 70-year-old their skin looks like a disaster area, full of wrinkles, moles, liver spots and scabs that are actually small basal-cell carcinomas. In the dermatology department we could do little more than freeze these basal-cell carcinomas with liquid nitrogen. A basal-cell carcinoma is a type of skin cancer that is far less dangerous than a melanoma because basal carcinomas seldom metastasise.

These two extremes illustrate the disastrous influence that sunlight has on the skin. Sunlight causes the skin to age faster. It is ironic that many people think that a tanned skin is healthy. Or at least that it makes them look healthy. But from the biochemical point of view, tanning is a molecular catastrophe. Tanning is after all the skin cells' defence mechanism against the DNA being damaged by the sun's radiation.

When you are lying on a beach in the sun enjoying a well-earned holiday, billions of bits of DNA in your skin cells are being damaged. The reason for this is that sunlight contains huge amounts of energy. Sunlight consists of UV light. When a beam of UV radiation hits the DNA in a skin cell, a chemical reaction occurs that creates 'thymine dimers'. This is what scientists call a mutation.

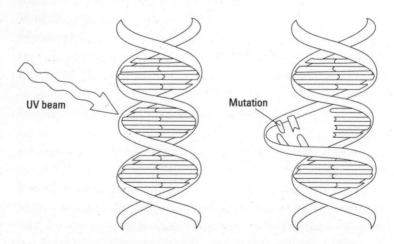

Sunlight (UV radiation) damages the DNA in the skin cells. The energy-rich UV light breaks a rung in the DNA ladder and will glue two thymine-molecules together, which deforms the DNA chain. This deformation is called a mutation.

Thymines are small molecules or pieces of DNA (a piece of DNA like this is called a 'base'). These pieces or bases together form the rungs of the ladder that we call DNA. Normally, a thymine is always connected to another base, called an adenine. A thymine and an adenine molecule together form one rung of the DNA ladder.

When a UV beam hits a thymine molecule it breaks away from the adenine molecule so that the rung is broken in two. Then the thymine will attach itself to a neighbouring thymine, thus forming a 'thymine dimer'. This will damage the DNA string, which is called a mutation. When you sit in the sun for a couple of hours, billions of thymine dimers will be created in billions of skin cells.

Skin cells will respond to this DNA damage by protecting themselves from sunlight. Every skin cell will protect the DNA in its cell nucleus by putting up a kind of umbrella of melanin granules above the cell nucleus. This umbrella that doesn't protect the cell nucleus from raindrops, but from UV radiation. Melanin granules are a dark substance and when each of the billions of skin cells build a defensive belt of melanin granules round their cell nuclei the skin generally goes darker. This is what we call tanning. Therefore, you go brown because the DNA in the skin cells is mutating due to the sunlight.

This is not good for us. The formation of a melanin shell round the cell nucleus is only a defence mechanism to limit any further damage to the DNA. A great deal of damage has already been done. Many billions of thymine dimers have already formed. Luckily the majority of these mutations are repaired by special proteins that rush to the DNA and check for mutations. Most of the mutations will be repaired, but not all.

Thus, the DNA in the skin cells will not be completely repaired, because there is no such thing as perfection in nature. The thymine-dimer story explains why the skin has a

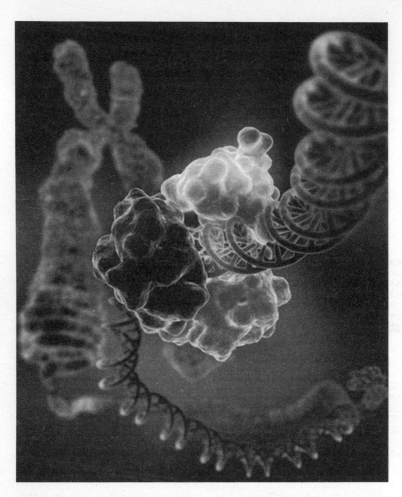

A protein covers a damaged DNA string to repair it.

memory: every exposure to the sun causes mutations to our DNA of which most (but not all) are repaired. In the long term, this damage will cause skin ageing and, if you are very unlucky, skin cancer. Studies have actually shown that children who have suffered from sunburn just three times in their younger years run four times the risk of developing skin cancer when they are older.[157] The skin never forgets. There exists a certain disorder in which the DNA mutations caused by sunlight are insufficiently repaired. This disease is called *xeroderma pigmentosum*. Every time people who have this disease are exposed to the sun, they will suffer DNA damage which will never be repaired. From a young age these people will develop skin tumours and skin lesions and their skin will age quickly with wrinkles, pigment spots and moles. If you are feeling brave, Google 'xeroderma pigmentosum' to see some shocking images. This disease is a good example of the damage done to DNA by sunlight when this damage is not repaired as completely as it is for the majority of people who have enjoyed a day by the seaside.

To be brief, sunlight is a major cause of skin ageing and wrinkles. This is why eating plenty of fruit is only one part of the story and why every dermatologist will insist that you protect your skin against the sun. So don't forget to use sun cream and wear a hat with a wide brim. Obviously we need sunlight to create vitamin D, but we can get vitamin D in the form of supplements. More on this subject later.

'Is fruit really good for us?'

Countless studies have shown that eating fruit is good for our skin, blood vessels and brains, and that it will even help us to live longer. In spite of all these scientific opinions and advice from doctors and scientists, I still, all too often, come across patients who have heard from somewhere that eating fruit is not good for us. They claim that fruit 'contains too much sugar' or that it can cause 'food allergies' or 'food intolerances'. There are even companies that offer food intolerance tests, which all of a sudden show that people are allergic to, or intolerant of, this or that kind of fruit. I think that this is a dangerous trend because even specialised laboratories in university hospitals find it difficult to identify certain allergies and intolerances. Even the biological characteristic of an allergy or intolerance – which is the presence of large amounts of immunoglobulin E-antibodies (IgE for allergies) or IgG antibodies (for an intolerance) to combat, for example, pollen, dust mites or a kiwi fruit – doesn't necessarily mean that you are allergic to or intolerant of pollen, dust mites or kiwi fruit. There are also people who have large amounts of antibodies in their blood against pollen or house mites yet have no allergies. So be critical of a less professional lab using tests that allegedly can determine whether or not you are allergic to or intolerant of an orange or a banana. People are often wrongly told that they are allergic to or intolerant of different types of fruit, so that they eat less fruit or even none at all, sometimes for their whole lives, which cannot be good. Moreover and ironically enough, eating fruit can actually be good for allergies. A study involving 1,500 asthma sufferers showed that people who eat just two apples a week had 32 percent less risk of an asthma attack.[158]

Then there's that other myth: fruit isn't good for us because it 'contains a lot of sugar'. But the glycaemic index of fruit is low. Most types of fruit don't cause high sugar peaks. Even a sweet apricot has a glycaemic index of only 15 (the glycaemic index of glucose is 100). This is because the sugars in the fruit are packed in fibre that slowly releases the sugars into the blood stream, which means that the insulin and IGF barely peak. This is opposed to a cake for example, which contains hardly any fibre at all and causes high insulin peaks. Additionally, besides fibre, fruit contains thousands of types of curative phytochemicals, healthy substances not found in cakes such as flavonoids, ellagic acid, coumarins, terpens, indoles and so on.

Then there's the misconception that you should peel your fruit first because of all the pesticides in the skins. Don't peel your fruit! There are so many healthy substances in and just under the skin. Of course there are pesticides on most fruits, but you can get rid of these by washing and drying the fruit thoroughly. The tiny amount of pesticides that might still remain are no match for the fibre and flavonoid rich peel. Plus, soft fruits such as strawberries, raspberries or bilberries can't be peeled anyway.

SUMMARY

A sufficient daily intake of **fruit**:
- reduces the average death rate;
- inhibits skin ageing (primarily with red and blue fruit, such as strawberries, raspberries, pomegranates and bilberries);
- keeps the blood vessels healthy;
- reduces the risk of Alzheimer's disease and other types of cognitive degeneration (this applies mainly to blue fruits such as bilberries, blackberries and myrtle berries).

Drink at least one **freshly squeezed fruit juice** every day. To minimize sugar peaks:
- add fibre to the juice;
- add vegetables (which contain less sugar than fruit).

Do not peel fruit but wash and dry it thoroughly.

The **formation of wrinkles** can be suppressed in the following ways:

From the inside:
- eat plenty of fruit, especially red and blue fruit;
- drink freshly squeezed fruit juice every day;
- drink green and white tea;
- limit your intake of milk products, red meat and sugar;
- stop smoking.

From the outside:
- use sun cream with at least factor 30, every two hours;
- wear a hat with a wide brim;
- use day cream every day with a solar protection factor (SPF) of at least 15 (4 is too low);
- if you really want a tan, then use tanning lotion.

Step 3: Fish, poultry, eggs, cheese, tofu and Quorn

- Fast food (hamburgers, pizzas, hotdogs, snacks)
- Red meat (pork, beef, lamb, horse)
- Fried food

replace with

- Fatty fish (salmon, mackerel, herring, anchovies, sardines), poultry (chicken and turkey), eggs, cheese, tofu, Quorn.

Fatty fish and poultry

As we saw in the chapter about proteins, too much meat can increase the risk of cancer, osteoporosis, accelerated ageing and heart disease. However, giving up meat is not really necessary because meat contains several important nutrients, such as zinc, iron, vitamin B12, creatine and carnosine. Thus, eating small amounts of meat can be good for us. Every day you could eat a piece of meat the same size as the circle formed by placing the top of your thumb against the top of your index finger. Try to avoid red meat, such as pork, beef or lamb, which do increase the risk of cancer, heart and vascular disease and diabetes; eat easily digestible white meat such as poultry and, better still, fatty fish that is full of omega-3 fatty acids. A study among 120,000 test subjects showed that each portion of red meat that is substituted by poultry reduces the death rate by 14 percent.[159]

People often think they need meat in order to give them enough proteins. We are always told that meat will make us grow strong and healthy. But surely, elephants and giraffes are exceptionally big and strong and they never eat meat. They get their proteins from plants. Humans should do the same. Vegetables and legumes are full of proteins. These proteins are better for us than animal proteins because they contain less of the protein-building amino acid, methionine. When we were discussing legumes, we said that a lower intake of methionine reduces the protein production, which made lab animals live longer. Vegetable proteins also contain less sulphur-containing amino acids. These sulphur-containing amino acids acidify the blood. To combat this acidification, calcium carbonate is absorbed from the skeleton. This weakens the skeleton and increases the risk of osteoporosis. People who eat too much animal protein will find that their bones deteriorate faster as the years go by than those who eat mainly vegetable proteins:

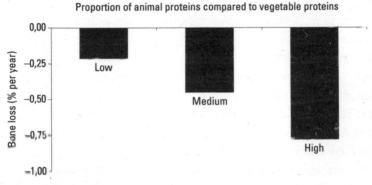

People who eat small amounts of animal proteins and large amounts of vegetable proteins (left column) have annually much less bone loss than those who eat large amounts of animal protein and small amounts of vegetable protein (right column). *Source: A high ratio of dietary animal to vegetable protein increases the rate of bone loss and the risk of fracture in postmenopausal women*, American Journal of Clinical Nutrition, 2001

Another reason why vegetable proteins are better for us than the proteins contained in meat is that the proteins in plants, unlike animal proteins, are not surrounded by unhealthy fats like inflammatory omega-6 fatty acids.

There are countless meat-substitute products that contain vegetable proteins, such as tofu (made from soya beans), legumes, Quorn (made from a fungus), and don't forget vegetables, like broccoli. Briefly, cutting down on meat doesn't mean that you will have to go short of proteins. However, it is advisable to eat small amounts of poultry and fish because they contain certain unique nutrients that we need. In this regard we should point out that the meat (and other products that come from animals such as eggs) that we eat today is not the same as the meat from roughly fifty years ago.

In the past, cows grazed in fields and all chickens were free-range. Today, cows and chickens are found in large breeding sheds where they are fed on corn, wheat and mixed cattle fodder. The meat we get from cows and chickens that are fed on this kind of diet is not as healthy for us. Research has shown that the bodies of those who eat non-grass-fed cows contain fifteen to forty times more inflammatory omega-6 fatty acids than omega-3 acids.[160] In an ideal world, the ratio between omega-6 fatty acids and omega-3 fatty acids should be less than four to one. The same applies to chicken and their eggs. A study appeared in *The New England Journal of Medicine* which showed that the eggs from corn-fed chickens had twenty times more omega-6 fatty acids than omega-3 fatty acids.[161] For free-range chickens, the ratio is one to one.

On top of this, cows, pigs and chickens are given antibiotics to prevent infections. This is also bad for our bodies, but in a different and indirect way. In particular, the antibiotics given to chickens are a cause for worry. Chickens are given preventive antibiotics and this creates super resistant bacteria in and on chickens; bacteria that ends up on our plates and from there, enter our gut. Nowadays, doctors continually have less and less effective antibiotics at their disposal to combat a particular type of bacteria; the gram-negative bacteria. These bacteria live mostly on the skins of the chickens and in human intestines. Certain super-resistant gram-negative bacteria have emerged and can be traced to those battery farms where the chickens are given large doses of antibiotics. If this carries on, it won't be long before doctors can only prescribe the highly toxic antibiotics to combat infection with gram-negative bacteria. In some cases these antibiotics no longer work at all. It's becoming more common for a patient to die from a resistant infection caused by super-resistant bacteria, in spite of all the most powerful antibiotics available in the twenty-first century.

The pity of it is that research shows that all this policy of preventive antibiotics is unnecessary. Young pigs that stay with their mothers for longer are just as resistant to infection as those given antibiotics.[162] If pigs stay with their mothers for longer, their immune systems can better develop. These days, young pigs are taken away from their mothers as soon as possible so they can be fattened up.

In short, the meat we eat today is not the same as it used to be. The same goes for eggs and dairy products. I recommend eating more organic meat and eggs. Organic products come from animals that are brought up on grass and that have the space and time to grow in a natural environment.

SUMMARY

Contrary to what goes for animal proteins, the following applies to **vegetable proteins**:
- they contain less sulphurous amino acids which acidify the blood;
- they contain less methionine which stimulates protein production;
- they reduce the risk of osteoporosis;

Continued overleaf

- they offer many other benefits for our health because they are ingested through vegetables, legumes and other nutrient-rich sources.

Chickens and cows no longer eat **grass**, so that their flesh and eggs contain more of the inflammatory **omega-6 fatty acids** than the inflammation-reducing omega-3 fatty acids.

Organic meat and eggs come from animals that grow in a more natural environment.

Eggs and cheese

It has often been claimed that eggs are not good for us because they contain lots of cholesterol. It is true that an egg does contain a lot of cholesterol, but research has shown that the amount of cholesterol in our bodies does not rise significantly due to eating a couple of eggs each week. At first sight, this is not so surprising given that most of the cholesterol in our bodies is made by our bodies themselves. So it doesn't matter that much if we eat small or large amounts of cholesterol-rich food, because our own bodies regulate our cholesterol levels. It also seems that, even if eggs would increase the cholesterol level, this isn't always such a bad thing. Contrary to popular belief, a high cholesterol level is not always a bad thing. In fact, in spite of all those 'low cholesterol' margarines, cheeses or crisps, the total cholesterol level has little influence on heart and vascular diseases. Many people have high cholesterol levels without being at risk of heart and vascular disease (such as in Japan). And there are people who have low cholesterol levels (such as in Israel, where devout Jews use mostly omega 6- rich margarine in place of butter) and still suffer with many heart and vascular problems. Unless you have a rare genetic disease that dramatically increases your cholesterol level, it is not the amount of cholesterol that matters but rather what happens to the cholesterol in the blood: if the cholesterol is saccharinated (glycated) due to high levels of sugar in the blood, then it becomes more sticky, making the cholesterol in the blood vessels stick more easily to the blood vessel wall and thus cause hardening of the arteries. Or if the cholesterol is oxidised because we don't eat enough fruit or vegetables, then it also becomes sticky, with the same consequences. So it's not so much the amount of cholesterol that matters, but what happens to the cholesterol in the blood.

To cut a long story short: you can eat eggs, to a certain extent.[163] A large study among 120,000 test subjects demonstrated that even people who ate an egg every day didn't run a greater risk of getting a heart attack or a stroke. However, people who suffer from diabetes are at risk when they eat an egg every day. Just to be on the safe side, don't eat too many eggs: a few eggs a week are more than enough. I would even opt for one or at most two eggs a week if you are male and older than fifty, since men over 50 who eat more than 2.5 eggs a week have an increased risk of getting prostate cancer. And it is best to eat soft-boiled or fried eggs, instead of an omelette. When you make an omelette you stir the egg yolk in the pan which exposes the cholesterol in the yolk to oxygen and oxydises it more, which is less healthy.

And what about cheese? We haven't heard anything positive about dairy products like milk and yoghurt. But cheese is an exception. Studies have shown that not all dairy is equal, and that cheese can even reduce the risk of a heart attack, while milk and yoghurt have no effect. And all this despite the fact that cheese contains large amounts of saturated fats. One reason is that cheese is a major source of vitamin K2 or menaquinone.[164] Vitamin K2 is a substance that has a favourable influence on the heart and blood vessels.

This is why cheese is the only dairy product to be included in the food hourglass. Cheese is perfect for those of us who want to eat (wholemeal) bread occasionally. We can put cheese on our bread instead of meat. What is interesting is that centenarians who live in the Blue Zones (areas in Japan, Sardinia or Costa Rica which have an unusually high number of people who live longer than one hundred years – sometimes five times more than anywhere else) often habitually only eat small amounts of meat but they do eat cheese like delicious goat's cheese for example.

SUMMARY

High cholesterol does not play such an important role for most people when it comes to heart and vascular disease because:
- it's our own bodies that make the largest amount of our cholesterol;
- what happens to the cholesterol in our blood is more important:
    - saccharinated (glycated) cholesterol (as a result of high sugar consumption or diabetes) is stickier;
    - oxidised cholesterol (caused by too few flavonoids from fruit, vegetables or green tea) is stickier.

Eating a few **eggs** per week does not influence the cholesterol level.

Cheese contains saturated fats but does not increase the risk of heart and vascular disease.
Although cheese is made from milk, it nonetheless can be consumed because cheese:
- is made from super-digested milk (digestive enzymes and bacteria are added to milk to make cheese)
- is an important source of vitamin K2 (menaquinone);
- contains bacteria that keep the gut healthy (probiotics, especially in old cheese).

Fried food and fast food (hamburgers, hotdogs,
pizzas, lasagne et cetera)
Everyone knows that fast food and fried foods are not good
for us; and everyone is quite right. First of all they contain
trans fats. It is primarily these 'unnatural' trans fats which
the body does not know what to do with, and they lump
together in the blood vessels, the brain and the fat cells. Not
only that, fast foods and fried foods raise the blood sugar
level, which increases the risk of diabetes, heart and vascular
disease and cancer. These foods cause such high glucose
peaks and also contain so many phosphates that they, as
we saw earlier, cause accelerated ageing. Also, fried food
and fast food hardly contain any vitamins and minerals, let
alone flavonoids or other healthy phytochemicals. In other
words, they are fake foods that are made in such a way that,
because of the large amounts of sugars, fats and salt they
arouse the primitive parts of the brain so that they cause
delight and addiction. Sugars and fats were foods seldom
seen in the African savannahs of our ancestors so our ancestors couldn't stuff themselves with these deliciously addictive
foods. Nowadays, however, we are surrounded by this kind
of food and this excess is causing enormous health problems.
Food manufacturers couldn't care less about this because
they continually lump sugars, fats and salts together in such
a way to design foods that give people a 'flavour orgasm'.
The result is a Mega-burger or the Big Whopper. With chips
of course, or other fried food. Even healthy food is rendered
unhealthy by throwing it into the frying pan, seen for
example in the connection between non-fried fish and heart
failure. Heart failure is an illness usually caused by a heart
attack. The patient survives the heart attack, but has to carry
on for several years with a damaged heart, and will in
most cases die from heart failure four or five years after the
original attack. It seems that eating fish five times per week
reduces the chance of heart failure by 30 percent. However,

if the fish were to be fried, then the chance of heart failure rises by at least 48 percent, even with just one portion of fried fish per week ...[165]

Of course, all that fast food and fried food is simply delicious. The average Englishman can't do without his weekly portion of fish and chips, whether or not accompanied by a battered sausage or hamburger. It may seem difficult to leave these foods behind. Bear in mind, however, that I often hear from people who haven't eaten fast food for some time say that they find it hard to digest fried foods when they go back to eating a fried meal or into a fast food restaurant. The food lies heavy in their stomachs, they can't sleep properly, the next day they are tired, pale or bad tempered. Their bodies are no longer used to digesting and processing such unhealthy foods. This is good, because it almost automatically ensures that people don't relapse into eating fast food on a regular basis.

SUMMARY

Fried foods and fast food:
- contain **trans fats** that clump together everywhere in the body, causing hardening of the arteries and micro-inflammation;
- cause **high blood sugar and insulin peaks**;
- contain **phosphates** which accelerate the ageing process;
- contain almost **no** flavonoids, vitamins, minerals or other nutrients;
- are **addictive** because of their composition (fat, salt and sugars);
- contain a large amount of **calories**.

Even healthy foods will be unhealthy if you fry them.

Step 4: Dark chocolate, nuts, soya dessert, soya yoghurt

> Sweets (biscuits, pastry, candy bars, ice cream)
>
> **replace with**
>
> Dark chocolate, nuts, soya dessert, soya yoghurt.

About sweet snacks

Human beings appear to be hungry all the time. Their three meals a day are not enough, they like to top these up with all sorts of snacks, four o'clock munchies, and late-night bites. But it's these snacks that creep into a diet and secretly but hugely increase the amount of calories. Polishing off three madeleine cakes (it just takes a minute while you're still glued to the television) will add up to almost 400 extra kilocalories, which is a sizeable part of the 2000 kilocalories a person requires each day. A praline contains 130 kilocalories. Eat four pralines a day and you have already consumed one quarter of your necessary daily portion of calories.

These snacks are not just calorie bombs, they are extremely addictive too. Each time you eat something sweet, the sugar in your snack will almost immediately find its way into the bloodstream (since the snack contains hardly any fibre or other substances that help ensure a slower absorption of sugars). This sugar peak will give you a kind of sugar rush: you feel content and satiated, and quite energetic. But what goes up has to come down at some point (unless it has reached earth's escape velocity) and this sugar peak will be followed by a sugar dip: a drop in the amount of glucose in the blood. That is because the sugar peak has been followed by an insulin peak that forces the sugar to be absorbed by the cells in the body, especially in the liver, muscle, and fat cells where the sugar is stored. This causes the sugar to disappear from the bloodstream. As a result, the amount of sugar in your bloodstream will decrease just as rapidly as it has increased. And lots of people exhibit withdrawal symptoms because of this. They feel light in the head, weak or tired, they have trouble concentrating, they have sudden hunger pangs or a strong urge to eat 'something sweet'. Some people even experience shaking legs and sweating. These are literally withdrawal symptoms; after all, sugar is addictive. Scientific research has even proven that sugar influences the endorphin system, just like heroin – endorphins are substances that make us feel good (junkies inject themselves with heroin because this substance activates the endorphin system). And sugar can indeed cause withdrawal symptoms, just like heroin or any other drug; that is why some psychiatrists are trying to include sugar as an addictive substance in the DSM (the DSM – Diagnostic and Statistical Manual of Mental Diseases – is the psychiatrist's bible and includes various addictive disorders as well). But a 'sugar addiction' will probably never be incorporated in the DSM, since food that is rich in sugar still remains food, and a human being needs food every day to survive, unlike drugs such as heroin or cocaine.

That does not alter the fact that sweets are addictive. If you eat two delicious marzipan cookies at 4pm every day for a few days in a row, most people will develop a craving for such a marzipan cookie at 4pm after just a short while. This isn't real hunger but an addiction to sugar-rich food. It is important to realise that this craving for sweets isn't 'genuine' hunger, but just a withdrawal symptom. Real hunger is coupled with sensations at the back of your throat and a desire to eat any food at all, no matter what. Such food doesn't necessarily need to be sweet. But nowadays, the hungry feelings experienced by many people are withdrawal symptoms caused by a craving for sweet things. This 'pseudo appetite' is not accompanied by sensations in your throat but by a strong craving for anything sweet. And by withdrawal symptoms such as a feeling of faintness, fatigue, wobbly legs, nervousness, concentration disorders or sweating.

The fact that sugar can be addictive makes it extra hard to stop eating sweet snacks. The only way to get rid of this habit is to tackle this problem the same way as any other addiction: stop it at once, and completely. Stop eating sweet snacks for ten days and the daily urge to eat something sweet will completely disappear. Of course, the first couple of days are the most difficult. To live through these first days you need to make sure that there are no sweets to be found anywhere in your house. This way, you can't get tempted by sweets when you are hungry (and desperate). So throw out all cookies, chocolates, pastry and candy along with the packets of potato crisps and other unhealthy items (crisps too consist of more than 50 percent of sugars and 'healthy' fat-free crisps are 80 percent carbohydrates). Once all this sweet stuff has disappeared from your home there is nothing you can do but eat a healthy snack. At this point you won't need to conquer your urge but just suffer through it. If your housemates don't agree to this massive clearing out of sweet stuff (because they still want to munch on sweets them-selves), make sure they have their own sweets cupboard; keep it locked and don't keep a key yourself.

But then what? Suppose it's 4pm and you are hungry. What healthy and tasty snacks can you eat? In the food hourglass, these snacks are placed in the fourth group. This is because many of these snacks constitute an important part of a healthy diet, such as nuts (walnuts, hazelnuts or almonds) and dark chocolate. Or you can eat a bowl of oatmeal porridge, soya yoghurt or soya dessert. And of course you can combine these foodstuffs with fresh or dried fruit.

For instance, my four o'clock snack consists of oatmeal porridge with chunks of apple in it; sometimes I just eat an apple with a piece of dark chocolate. Really delicious! (Imagine, in the past I didn't even like dark chocolate.)

Here are a few other tips for an ideal four o'clock snack:

- a bunch of grapes and a peach;
- dark chocolate with strawberries;
- some walnuts with blueberries;
- caramel-flavoured soya dessert with chunks of walnut in it, and a spoonful of linseed (rich in omega-3);
- an avocado (which you cut in halves which you can spoon out. You can sprinkle a bit of olive oil, salt and pepper over it);
- dried fruits with nuts;
- olives and feta cheese;
- cherry tomatoes, cauliflower and broccoli flowerets that you can dip in hummus (mashed chicken peas with olive oil, salt and garlic), pesto (made from olive oil, basil, nuts and cheese) or guacamole (an avocado based dip). You can buy these dips in the supermarket.

People who substitute their daily portion of sweets with these healthy snacks can experience withdrawal symptoms in the first couple of days. But afterwards, their body will adapt to this healthy food. As a result, they will experience genuine feelings of hunger again, and not a pseudo appetite coupled with a strong craving for sweet things, dizzy spells, shakes, etc. Furthermore, their taste buds will change which will cause 'sweet' to taste differently. In the beginning, just after you have stopped eating sweet snacks, grapes and straw-berries will taste like 'nothing', especially compared to chocolates or a slice of cake. But after just a few days, your taste buds start changing; grapes, strawberries and other fruits will taste much sweeter. After a while, lots of people will find that cakes or pastry taste far too sweet. Some may even find it 'awful' because they are so sweet. Fruits and other food will taste more refined and sharper. I would say: just try it, and you will be amazed at the ways in which your body manages to adapt, ranging from your daily late afternoon fainting spells that will disappear, to your sense of taste that will change as well.

SUMMARY

Snacks greatly increase the number of daily calories.

Sweet snacks are highly addictive and cause **withdrawal symptoms (pseudo appetite),** such as:
- a sudden strong craving for sweet things;
- fatigue;
- concentration disorders;
- sweatiness;
- wobbly or weak legs.

These withdrawal symptoms are often mistaken for real hunger. But **genuine hunger** is characterised by:
- an ordinary (not strong) craving for food (no sweets);
- an absence of withdrawal symptoms (see above);
- a kind of 'sense of taste' at the back of the throat.

The only way to **conquer** this daily craving for sweet things is:
- total abstinence: get rid of all sweet snacks and throw them out of the house (cookies, cake, ice cream, but also crisps) and replace them with healthy four o'clock snacks, such as:
    - (re-heated) oatmeal porridge;
    - (dried) fruits such as grapes, peaches, strawberries, blueberries, apples, etc.;
    - soya dessert or soya yoghurt;

Continued overleaf

- dark chocolate;
- nuts.
- also: olives, cheese, avocado, cherry tomatoes, cauliflower, broccoli flowerets, paprikas, on their own or with hummus and other dipping sauces.

After a while, the withdrawal symptoms (or pseudo appetite) will disappear and your sense of taste will change (fruit will taste really sweet again).

Dark chocolate

Dark chocolate is chocolate that contains more than 70 per cent cocoa. This chocolate has unique effects because it contains large concentrations of a special type of flavonoid. As we have seen before, flavonoids are also present in fruit (they contribute to the bright colours of the fruit) and in green tea. Cocoa contains even more (and more powerful) flavonoids than green tea. Dark chocolate has a beneficial effect on the cardiovascular system, it lowers blood pressure and it makes blood platelets less sticky.[166-167] Moreover, dark chocolate engenders a better insulin response. This means that the body becomes better at handling and storing sugar, and also the sugar will stay in the bloodstream for a shorter period, so it cannot reach all kinds of tissue where it can cause damage, such as the eyes, the brain, the kidneys or blood vessel walls. Studies have shown that 'pre-diabetics' (people whose blood sugar levels are out of control but not so much as to cause actual diabetes) who eat a piece of dark chocolate every day over a period of 15 days, exhibit an insulin response that is twice as strong as before. Furthermore, their diastolic blood pressure goes down by nearly five points, which is a better result than various types of blood pressure medication can achieve. Especially because it is often difficult to lower the blood pressure of people who suffer from (pre-) diabetes.

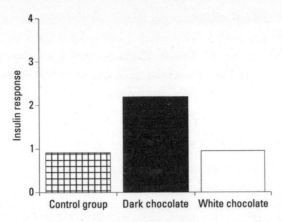

The dark bar represents the insulin response (IR) for the people who were fed dark chocolate. The white bar represents the IR of the control group who were fed white chocolate. The chequered bar represents the IR prior to the start of the study. The higher the IR, the better, because this means the sugar will leave the bloodstream quicker.

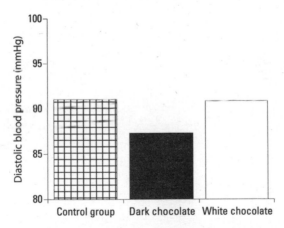

The blood pressure of the people who were fed dark chocolate will decrease (black bar). The blood pressure of persons who ate only white chocolate will remain the same (white bar). The chequered bar represents the blood pressure prior to the start of the study. *Source: Blood pressure is reduced and insulin sensitivity increased in glucose-intolerant, hypertensive subjects after 15 days of consuming high-polyphenol dark chocolate.* Journal of Nutrition, 2008

But dark chocolate, and especially its main ingredient, cocoa, has more good things to offer. Let's take a look at a study conducted among the native Kuna tribe, living on a group of islands off the Panama coast. The Kuna drink large quantities of warm cocoa powder, up to four or five mugs a day. Since cocoa contains a lot of flavonoids, scientists consider the Kuna to be 'the people with possibly the most extensive flavonoid-packed diet in the world'.[168]

The Kuna are especially well-known to the medical world because high blood pressure (hypertension) barely exists among this population. On the other hand, in the West, hypertension is a disease that affects 30 percent of the population. And the older you become, the more chance you have of getting high blood pressure. But even most elderly Kuna have normal blood pressure. The average blood pressure among these elderly people is 110/70, which is very low (according to the doctors, blood pressure higher than 140/90 is considered high; normal blood pressure is 120/80). At first, the scientists thought that the low blood pressure of the Kuna people was a result of good genes, but they couldn't find any genetic evidence for this. Research that was conducted later on demonstrated that the Kuna who moved from their islands to the Panama mainland, have started to suffer much more from hypertension. The Kuna who have moved to the mainland were mainly on a western food diet. The scientists were highly amazed when they examined the 78,000 death certificates of the Kuna living on the islands and of the Kuna who had moved to the mainland. The scientists concluded that death by cardiovascular diseases occurred nine times less among the Kuna living on the islands, compared to the Kuna living on the mainland. Cancer was found 16 times less, and diabetes occurred four times less than on the mainland.[168] Also, the number of people older than 75 was 2.3 times higher on the islands than on the mainland.

Of course, these developments are not just due to the cocoa. The Kuna on the islands ate a lot more vegetables, fruit and fish, compared to the Kuna on the mainland, who mainly adhered to a western diet. But this study does demonstrate the importance of a healthy diet. The fact that the Kuna had nine times less chance of suffering a heart attack had nothing to do with their genes, but was due to their way of life.

However, lots of people will voice their dislike of dark chocolate. This applied to me too, at least a while ago. But after having read numerous studies on dark chocolate, I decided to try and get used to eating it. Frankly, the first time I ate a piece of 86 percent dark chocolate, it tasted awful, to say the least. Real dark chocolate is very bitter and I don't like bitter tastes at all. But I thought it was remarkable that after twenty minutes my head felt strange and I could concentrate a lot better, especially due to the massive vaso-dilatation (among other things, dark chocolate causes the blood vessels to dilate, which will result in lower blood pressure). In the following days I kept eating a little piece of this disgustingly bitter dark chocolate. As a result, by now I really like dark chocolate very much, and actually I can't do without it. You don't need to eat a lot of dark chocolate to benefit from its healthy effects. Ten grams of dark chocolate (about one-fifth of a chocolate bar) suffices to reduce your risk of a heart attack.

For those who cannot overcome the barrier of eating dark chocolate, brown chocolate may be an alternative, provided that it is taken with a certain measure. A recent meta-analysis study, published in *British Medical Journal* and including 114.000 test subjects, showed that people who regularly eat dark or brown chocolate, have a 37 percent decreased risk of getting a heart attack.[169]

SUMMARY

Dark chocolate (more than 70 percent cocoa):
- contains more (and more powerful) flavonoids than green tea;
- lowers blood pressure;
- enhances the insulin sensitivity of the body (sugars stay in the blood stream for a shorter period);
- reduces the chance of getting cardiovascular diseases.

A daily diet of large doses of pure **cocoa** (as is the custom with the Kuna tribe), in combination with a healthy life style, dramatically reduces the chances of getting cardiovascular diseases, cancer, and diabetes.

And for those who don't like bitter tastes, you can actually learn to eat dark chocolate.

Soya dessert and soya yoghurt (and fibre!)

Soya can be used for making porridge too. You can also buy soya dessert or soya yoghurt in most supermarkets. Soya desserts are sold in most supermarkets, in various flavours, such as chocolate, vanilla, caramel, etc. Of course, these desserts often contain quite a lot of sugar. You can reduce the sugar peak by adding a spoonful of fibre to this kind of dessert (oat bran, for instance). The fibre makes sure that the sugar in the dessert isn't absorbed into the bloodstream so fast. Furthermore, the fibre improves the bowel transit and keeps the gut healthy. You can also add chunks of walnut or a spoonful of linseed oil to the soya dessert, which will add extra fat to the dessert. Fats slow down the stomach-emptying, which means that the sugar in the soya dessert doesn't reach the bowels as quickly and the sugar peaks are less high. Or you can make your own soya dessert or porridge and replace the sugar with stevia. It goes without saying that these tips also apply to soya yoghurt.

I just mentioned fibre; fibre is very important for our

health. It keeps your gut healthy, and there is no such thing as a healthy person with an unhealthy gut. If the gut is ill, eventually we will become ill too. Fibre also has a big influence on the composition of the gut bacteria. Our gut forms a huge habitat for bacteria: more than a hundred thousand billion. In your bowels we find more than ten times as many bacteria as there are cells in your body. Everyone has a different composition of bacteria in his or her gut. A recent study in *Nature* even found that there exist three different bacterial ecosystems, called enterotypes.[170] Just like with the four different blood types, people can also have one of three different intestinal bacterial groups. This bacterial composition influences our health. For example, people who have enterotype 1 have many intestinal bacteria that produce vitamin B2, B7 and vitamin C. People with a different bacterial make-up can have a predisposition for getting fatter, or run a greater risk of getting bowel cancer.[171] In short, it's always important to have a healthy population of gut bacteria. A healthy diet, with lots of fibre, ingested through fruit and vegetables, and few dairy products, red meat and fast sugars can provide a healthy habitat for the bacteria in your gut. Fibre can directly influence the gut bacteria.[172] That is why I recommend eating one to three teaspoons a day of insoluble fibre foods, such as oat bran, which you can find in every organic food store ('insoluble' means that the fibres don't dissolve in water). This fibre can be added to soya dessert, fruit juice, vegetable juice or soup. The soluble fibre is important as well, and can be taken in the same quantities as the insoluble fibre, that is to say, one to three teaspoons of soluble fibre a day. Some food manufacturers offer 'multi-fibre' compounds that contain both soluble and insoluble fibre, sold in large jars. Especially soluble fibres such as inulin, Arabian gum, and oligofructose (fructo-oligosaccharides) stimulate the growth of healthy gut bacteria and improve the bowel movement and various intestinal complaints.[173–174]

The gut bacteria maintain a delicate balance that continuously influences our health. This balance is easily disrupted, not only by unhealthy eating habits, but also by such things as antibiotics, for example. Antibiotics are weapons of mass destruction for our gut bacteria. That is why antibiotics can have various side effects. For instance, antibiotics will see to it that less vitamin K is produced, since gut bacteria are responsible for the production of vitamin K. Vitamin K plays an important part in the coagulation of the blood and in cardiovascular diseases. In this way, using antibiotics for more than ten days in a row can cause a vitamin K deficiency. Antibiotics can also be the cause of diarrhoea, because of the massive extinction of gut bacteria, and because the perished healthy gut bacteria are being replaced by unhealthy, resistant bacteria. And because so many benign bacteria are disappearing, fungi can also take their place. These fungi will proliferate within the bowel system, but also in the mouth, the vagina or the anus; that is to say, in all the areas that are usually occupied by harmless bacteria.

As long as antibiotics are prescribed and ingested for the right symptoms, I've got nothing against using them. But you shouldn't use antibiotics to cure a common cold or bronchitis, which are caused by viruses more than 90 percent of the time; and antibiotics won't destroy viruses. Nowadays, far too many people still rely on antibiotics to cure their cold, and in this way they not only pave the way for bacterial resistance but they also re-program their entire gut flora.

SUMMARY

You can add the following items to **soya dessert** and **soya yoghurt**:
- a teaspoon of fibre;
- chunks of walnut or other types of nuts;
- a teaspoon of linseed.

Fibre:
- takes care of a better bowel peristalsis;
- influences the composition of gut bacteria, and because of this also determines which vitamins, short chain fatty acids, and other nutrients are produced by the intestinal bacteria. These substances are of great importance to our general health;
- lowers the sugar peaks in the blood because it 'envelops' sugar molecules;
- can be added to soya dessert and soya yoghurt, but also to soup, fruit juice or vegetable juice.

There are 2 types of fibre:
- insoluble (in water) fibre (such as oat bran);
- soluble (in water) fibre (such as inulin, Arabian gum, and oligofructosis).

Consume 1 to 3 teaspoons of soluble fibre, and 1 to 3 teaspoons of insoluble fibre a day.

Nuts

Finally, we have arrived at the nuts! Nuts are very interesting foodstuffs. They contain lots of fats and that is why people often say nuts will make you fat, which is nonsense of course. It's evident that nuts are loaded with fats, a nut consists of more than 50 percent fat. But we have already discovered that not all fats are alike. Nuts consist of very healthy fats. Walnuts are an important vegetable source of omega-3 fatty acids. Fish contains omega-3 fatty acids such as EPA and DHA, while walnuts contain a third type of omega-3 fatty acid, namely alpha-linolenic acid. Omega-3 fatty acids are known to make people lose weight instead of getting fatter. This may seem strange because, strictly speaking, 100 grams of fat contain about 900 kilocalories, which is quite a lot. But this has been measured by burning 100 grams of fat in a glass jar in a laboratory, where the resulting release of energy is measured (in kilocalories). However, the human body is not a glass lab jar, and it has different ways of dealing with the various types of fat it encounters. In short, the fats in walnuts are very different from the 'fattening' fats in a burger, or in fries.

Because walnuts contain so many omega-3 fatty acids they are also called 'brain food'. As we have seen earlier, omega-3 fatty acids play an important part in the build-up and functioning of the brain. For example, research has demonstrated that Alzheimer-infected mice that are put on a diet of lots of walnuts deteriorate slower, in terms of their memory and learning ability, than mice that aren't fed walnuts.[175]

But keeping the brain healthy is not the only thing that nuts do. Extensive studies demonstrate that nuts have a big influence on the heart and vascular system. Women who eat a handful of walnuts every day, have 45 percent less chance of contracting a cardiovascular disease, according to the Nurses' Health Study, one of the most well-known studies ever conducted.[176] Indeed, the conclusion of the researchers was: 'Given the strong scientific evidence for the beneficial effects of nuts, it seems justifiable to move nuts to a more prominent place in the United States Department of Agriculture Food Guide Pyramid.'

They didn't really act upon this recommendation, since the US Department of Agriculture has recently abandoned the food pyramid in favour of a food plate that consists of just four large sections (fruits, vegetables, grains, proteins, and a separate section for dairy products). Nowhere are nuts shown. Of course the main complaint is that this new food diagram is much too simple. And that's certainly true! This food plate is ridiculously oversimplified (since it is meant for the 'general public') and more than half of it consists of precisely those products we should consume less.

But let's get back to the nuts. Walnuts in particular are very beneficial to the heart and vascular system. Not only because walnuts contain omega-3 fatty acids, but because they also contain arginine, an amino acid that dilates the blood vessels and keeps the vascular walls healthy. Walnuts also contain different types of vitamin E. In the natural world, eight types of vitamin E exist. But most dietary supplements contain just a single type of vitamin E, called alfa-tocoferol. This is fairly absurd, since the body doesn't need just one type of vitamin E, but multiple types. Walnuts, on the other hand also contain gamma-tocoferol, among other things. This is a type of vitamin E that is said to be beneficial to the blood vessels, unlike alfa-tocoferol, which is frequently used in dietary supplements.[177]

As a result of all these substances in nuts, the blood vessels will be less susceptible to cholesterol precipitation. People who eat almonds, for example, have fewer oxidised cholesterol particles in their blood.[178] It is mainly the oxidised cholesterol particles that are sticky and cause a hardening of the arteries. Other studies have demonstrated that people who eat walnuts have 19 percent fewer ICAM proteins on their vascular walls. ICAM proteins are a kind of 'harpoon' that catch white blood cells and remove them from the bloodstream, where they can stick to the vascular walls and cause inflammation, which in turn will cause the blood vessels to clog even faster (in this way you could regard hardening of the arteries as a kind of inflammatory disease).[179]

SUMMARY

Eating a **handful of walnuts** every day considerably reduces the chance of getting a heart attack.

Walnuts are beneficial to the **heart and vascular system** because they contain the following substances, among other things:
– omega-3 fatty acids that make sure the blood vessels are less susceptible to inflammation;

Continued overleaf

- several types of vitamin E that make sure the cholesterol particles are less sticky;
- the amino acid called arginine, which widens the blood vessels.

Furthermore, studies have proved that walnuts can slow down the ageing of the **brain**.

Step 5: Sugar substitutes, healthy oils and flavour enhancers

---

- Sugar
- Salt
- Oils rich in omega-6 (corn oil, sunflower oil, palm oil, sesame oil), margarine, butter, high-fat sauces

replace with

- Sugar substitutes (stevia, tagatose, sugar alcohols, fruit)
- Healthy flavour enhancers (herbs, garlic, onion, lemon juice, vinegar, potassium)
- Healthy oils (olive oil, linseed oil, walnut oil, rapeseed oil, soya oil, perilla oil)

Oils, butter, margarine, and high-fat sauces
Which is healthier: margarine or butter? Neither one of them, but actually, margarine is even unhealthier than butter. Which seems strange, at first glance, because haven't we seen lots of promotion for margarine in the last couple of decades as a healthy alternative to butter? After all, margarine contains healthy vegetable fats while butter consists of animal fats.

Nevertheless, margarine is unhealthier than butter, because the vegetable fats in margarine mainly consist of inflammation-promoting omega-6 fatty acids. And not of anti-inflammatory omega-3 fatty acids. On the other hand we could say that this omega-6-filled margarine reduces the cholesterol level (a quality that is highly advertised by margarine manufacturers), but as we have previously stated in this book, cholesterol levels don't play a major part in cardiovascular diseases. What is important though, is what happens to the cholesterol: is it oxidised by too many free radicals, or is it 'coated in sugar' by excessively high levels of glucose in the blood, which renders the cholesterol sticky and dangerous. This also explains the 'Israeli paradox'. In Israel, religious Jews don't use butter with their meals, due to their religious laws. They use vegetable margarine as a substitute for butter. Because of this, Israeli people have very low cholesterol levels, in comparison to people in other western countries, but their scores are very high when it comes to cardiovascular diseases.[180]

In a word, it's not so much the total cholesterol level that counts, but it is the proportion of omega-6 to omega-3 that is important when we consider cases like the Israeli paradox. And in margarine this proportion is skewed.

But margarine manufacturers too have employees who occupy themselves with keeping up with the latest scientific research. And they have also read that omega-3 fatty acids are healthy. So currently we can also buy margarine 'containing omega-3 fatty acids' which is printed on the wrapping in large letters. But if you take a closer look at the list of ingredients, you will find that 100g of margarine contains 17g of omega-6 fatty acids, and only 3g of omega-3 fatty acids. Which means that the omega-3-'enriched' margarine still contains five times more unhealthy omega-6 fatty acids than omega-3 fatty acids . . .

Other manufacturers will zealously advertise the fact that their margarine is a source of healthy 'essential fatty acids'.

Essential fatty acids are fatty acids that cannot be produced by the body itself. Sounds good, a tub of margarine full of essential substances. But here's the catch: essential fatty acids can be both omega-3 and omega-6 fatty acids. Just like omega-3, the body cannot produce omega-6 fatty acids either, but this is not a problem because we ingest far too many of these fatty acids due to our unhealthy western eating habits. This way, manufacturers make their products appear healthy because they are said to contain 'essential' fatty acids. They are certainly essential, but also pro-inflammatory, and are present in our foodstuffs in far too large quantities.

From a scientific point of view, the claim that margarine lowers cholesterol levels 'coupled with a healthy diet and exercise' is really hilarious. The last sentence is often added in small print on advertisements. Of course such a statement is utterly worthless scientifically speaking, because the cholesterol level may be lowered mainly due to the extra exercise, or the healthy diet, and not by eating the margarine. Just imagine a scientific publication with this kind of title: 'Standing on your head twice a day will lower your cholesterol level, coupled with a healthy diet and exercise.' The article would even be absolutely truthful, but not a likely candidate for being published in *Nature*.

In the food hourglass, using margarine and butter is discouraged. Spreading butter or margarine on your bread is typically a western habit that is completely unnecessary. Also, lots of high-fat sauces such as tartar sauce or mayonnaise contain large doses of pro-inflammatory omega-6-rich oils, just like margarine. Rich omega-6 oils that are frequently used in the kitchen are corn oil and sunflower oil. That is why you should instead use healthy oils, such as olive oil, linseed oil, rapeseed oil, walnut oil, and possibly soya oil. These oils consist of a more balanced proportion of omega-6 and omega-3 fatty acids, especially compared to the omega-6-rich oils such as sunflower oil or corn oil.

Oil	Ratio omega-6: omega-3
Good ratio	
Perilla	1:5
Linseed	1:3
Rapeseed (Canola oil)	2:1
Walnut	5:1
Soya (bean)	7:1
Less good ratio	
Sunflower	30:1
Palm	46:1
Corn	83:1
Sesame	137:1
Arachid oil (peanut), Coconut oil	Doesn't contain omega-3 fatty acids

Some readers will have noticed that walnut oil and soya oil still contain relatively high quantities of omega-6 fatty acids in relation to omega-3 fatty acids (a bit like the current 'omega-3-rich' margarine that still contains much more omega-6 fatty acids). But the omega-6:omega-3 ratio in these oils is still superior to that of corn oil or sunflower oil (which contain dozens of times more omega-6 than omega-3 fatty acids). Moreover, these oils contain all sorts of phyto-chemicals that can be beneficial to the body, unlike margarine.

Olive oil is a special case. It isn't a rich source of poly-unsaturated omega-3 fatty acids, but it contains mainly monounsaturated fatty acids. These are fatty acids whose carbon chain just has one double bond. Olive oil is often promoted as being able to protect the heart, because olive oil is said to lower the cholesterol level, or because olive oil is an important ingredient in the Mediterranean diet that is protective of the heart. But this is not a basis for assuming that olive oil actually protects the heart. For instance, a high cholesterol level is not an important factor in cardiovascular diseases, and the Mediterranean diet may reduce the chances of getting a heart attack for totally different reasons than the olive oil (for example, because this diet contains more nuts, vegetables, and fruit). Recent studies have shown that olive oil doesn't seem to have a protective effect regarding cardiovascular diseases, contrary to the popular belief that olive oil is good for the heart.[181] Anyway, at least olive oil does not *increase* the risk of contracting cardiovascular diseases, which in itself is a good thing, since most oils used in the kitchen do increase this risk. However, olive oil appears to slow down various ageing processes, in the skin[182] as well as in the brain. For example, olive oil contains *oleocanthal*, a substance that slows the clumping together of proteins in Alzheimer's disease.[183] And oleocanthal doesn't do this because it is an antioxidant, but because it reacts in a specific way with the clumping proteins in the brain, and because it is anti-inflammatory. An article in *Nature* demonstrated that people who swallow a daily dose of 4 spoons of extra virgin olive oil ingest an anti-inflammatory dose that equals 10 percent of the powerful anti-inflammatory drug, Ibuprofen.[184]

Which types of oil are healthy to use while cooking? It is best not to cook with oils that contain many poly-unsaturated fatty acids (such as linseed, walnut or soya oil), because the double bindings in these oils are transformed into less healthy substances by the heat in the pan. Oils that are healthy to cook with are olive oil, Canola/rapeseed oil, and even better: avocado oil. It would be ideal to alternate between these types of oils.

SUMMARY

Butter is not recommended because it contains lots of **animal saturated fatty acids**.

Margarine is unhealthy because it contains trans fats (however, not all margarines contain them), and pro-inflammatory **omega-6 fatty acids**.

Olive oil is healthy because it contains **monounsaturated** fatty acids, and substances (**phenols**) such as oleocanthal:
- olive oil reduces cholesterol levels and forms an important part of the Mediterranean diet, but recent studies seem to indicate that it does not protect the heart after all;
- olive oil can slow down ageing processes in the brain (dementia) and the skin (wrinkles).

Healthy oils are: olive oil, linseed oil, walnut oil, Canola/rapeseed oil, perilla oil, and possibly soya oil.

Unhealthy omega-6-rich oils are: corn oil and sunflower oil, and the products that consist for the major part of these products, such as mayonnaise.

Healthy oils you can use to cook with are olive oil, avocado oil or Canola/rapeseed oil.

Sugar and sugar substitutes (stevia, tagatose, sugar-alcohols, fruit)

We use sugar in everything: in coffee and tea, to make jam, to sweeten our porridge, in our yoghurt or on our pancakes.

In this book we have already often discussed the influence of sugar on the body, ranging from a higher risk of getting cancers (that have a very high sugar metabolism) to acclerated ageing (through sugar that forms crosslinks between the proteins in our skin, eyes, kidneys, brains, etc.). The food hourglass recommends cutting down on sugar, just like almost every other food expert. But, are there any healthy alternatives to using sugar?

The best-known substitute for sugar is the artificial sweetener, aspartame. Aspartame is used for sweetening coffee and tea, but is also used in thousands of food products, such as soft drinks, cookies, and other sweet foodstuffs. However, there are quite a lot of health experts advising against aspartame, especially experts from the more 'alternative' side. It is often said that aspartame is carcinogenic,[185] or that it can disrupt the balance of neurotransmitters in the brain[186] (with a bigger chance of getting migraine attacks or depression).[187-188] But this effect seems to be much less serious than people initially assumed. Aspartame is said to be carcinogenic only in rats that ingest very high doses, doses that a human being would never consume. What is more important, however, is the negative effect that artificial sweeteners, such as aspartame, have on people's weight. We have already discussed this in the section on diet soft drinks that contain lots of artificial sweeteners. Artificial sweeteners, such as aspartame, activate all sorts of neurological and metabolic mechanisms in the body, which tend to make us fatter and run a greater risk of getting high blood pressure or diabetes. Artificial sweeteners appear to have a much more powerful effect on these mechanisms than non-artifical sweeteners like ordinary sugar.[189]

There are some healthy, natural sugar substitutes as well, such as stevia. Stevia derives from the stevia plant that grows in South America. Stevia is 30 to 100 times as sweet as regular sugar. Although it's a very sweet substance, stevia helps to stabilise the blood sugar and lower the blood pressure.[190] Stevia has already been used in South America for 1,500 years, and for 20 years in Japan, but it was not on the market in Europe until 2011, with the exception of France (according to professor Jan Geuns, biologist and researcher at the Catholic University of Leuven, the European sugar companies can be blamed for this, because they strongly lobbied to keep stevia off the market). But now stevia can be bought anywhere in Europe.

In supermarkets you can buy stevia that looks like ordinary sugar, that is to say, in the form of loose white powder or in the form of cubes. Although, when you take a look at the ingredients listed on the packet, you will see that this 'stevia' only contains 4 percent of stevia and 96 percent of dextrose (which is just another name for glucose). As already mentioned, stevia can be 100 times as sweet as ordinary sugar: this means you just need very small doses. That is why manufacturers add extra bulk to increase the mass, in the form of dextrose, which results in stevia powder or stevia cubes. But this way, you will still be eating mainly sugar. It is possible to buy pure stevia: it is available in liquid form, in small bottles. Just a few drops suffice to sweeten a dish.

Another healthy sweet alternative is called *tagatose*. Tagatose is a kind of natural sugar that tastes just as sweet as sugar, but the bowel absorption of tagatose is very small. As a result, tagatose hardly causes any rise in the blood sugar levels.

Sugar alcohols can also be a healthy alternative to sugar. Sugar alcohols (like xilytol, mannitol or erythritol; their name always ends in –ol) are only marginally absorbed by the gut, so they cause very small sugar peaks. Because so little of the sugar alcohols is absorbed they also contain far fewer calories. A healthy sugar alcohol is erythritol, which compared to many other sugar alcohols contains nearly zero calories and tends not to upset the gut. Some stevia powders in the store contain stevia and a sugar alcohol. So these stevia powders are ok.

A natural sweetener that is often mentioned as a healthy alternative to sugar is agave nectar (also called agave syrup). This substance looks a bit like honey and is made from cactus juice. Agave nectar is three times sweeter than sugar, but causes much weaker sugar and insulin spikes than regular sugar or honey. Sounds great, but still, agave nectar mainly consists of glucose, and especially fructose, which eventually are fully absorbed by the body. The fructose is mainly processed by the liver without causing high sugar spikes, however, this puts a large burden on the liver. That is why agave nectar is not recommended. The same goes for (raw) honey or maple syrup. Of course, if you already adhere to a healthy diet, eating small quantities of these foodstuffs every once in a while won't do much harm.

Another alternative for sugar is ... fruit. You can always sweeten your food by using apple sauce or mashed bananas, for example. Or you can use date sugar: this consists of mashed and ground dates.

One final remark: be careful with sweet things, even with healthy sweeteners. Using too many healthy sweeteners will sustain your addiction to sweets and make you eat more of other unhealthy sweet stuffs.

SUMMARY

Aspartame and other artificial sweeteners are advised against because they make you gain weight faster, even though they contain hardly any calories.

Healthy sugar substitutes are:
- **stevia**: derives from the stevia plant. Stevia is 30 to 100 times as sweet as sugar;
- **tagatose**: a natural sugar that is hardly absorbed by the intestines;
- **sugar alchohols** such as **erythritol** contains very few calories and causes very low sugar spikes;
- **fruit**: apple sauce, mashed bananas and date sugar are healthy sweeteners.

Agave nectar, (raw) honey, and maple syrup mainly consist of sugars (fructose, glucose and sucrose) and it is best not to use them too much.

Salt and potassium

It is often said – let's say almost always – that salt causes high blood pressure. And we all know that high blood pressure is unhealthy: the vascular walls become stiffer and high blood pressure 'pushes' the cholesterol particles into the vascular walls more easily. Also, the heart will need to pump harder in order to let the blood flow, which causes the cardiac muscle to swell and which can even cause heart failure in severe cases. And, of course, the higher the blood pressure, the bigger the risk is of a blood vessel bursting somewhere in the body. If this happens in the brain, doctors call it a stroke. If this happens in the belly it is called a ruptured aorta.

All in all, salt, hypertension, and cardiovascular disease appear to be linked together inseparably. But is salt actually as harmful as we think? An extensive study demonstrated that reducing salt-intake doesn't seem to play an important part in preventing cardiovascular diseases.[191] According to this study, a healthy person should even eat a bit more salt, since people in the study who ate more salt had less chance of contracting a cardiovascular disease, contrary to the general opinion. Keep in mind that this study needs to be interpreted very carefully. The researchers who conducted this study emphasise that the consumption of salt by people who are *already* suffering from some or other cardiovascular disease needs to be reduced. But other studies too demonstrate that reducing the consumption of salt doesn't have a pre-emptive effect on cardiovascular diseases in healthy people.[192] These studies have been widely publicised and have persuaded lots of people not to hold back when they cover their potatoes in salt. Although they are conveniently forgetting that other studies show that eating too much salt will increase the risk of getting a stroke by 23 percent.[193] In the medical profession it's important to keep in mind the greater picture at all times.

Why is it that people always used to say that salt is so bad for your blood vessels? The reason is that salt does indeed increase blood pressure (a bit). And in turn, lots of other studies will demonstrate that high blood pressure is bad for the heart and blood vessels. But, as we have previously seen in this book, this doesn't automatically lead to the conclusion that salt is bad for the blood vessels. It is not true that if a causes b, and b causes c, a will automatically be the cause of c as well. Medicine doesn't work like this. Salt increases blood pressure, and high blood pressure can damage blood vessels, but that doesn't mean that salt in itself is harmful to our blood vessels. According to some scientists, salt can also have a beneficial effect on the insulin response, or on the sympathetic nervous system, which can reduce the chances of contracting cardiovascular diseases.

However, what is very clear is that the intake of salt versus the intake of potassium is completely out of proportion in the West. We consume too much salt and too little potassium. In the body, these substances keep each other balanced. Maybe the growing number of people with hypertension in the West is not so much caused by eating too much salt, but by ingesting too little potassium. Potassium is mainly found in fruit and vegetables. Our distant ancestors in the Stone Age were on a diet that was mainly low in salt and rich in potassium. Nowadays it's just the other way round.

The intake of potassium can be dramatically improved by eating lots of vegetables and fruit. Notably, bananas, avocado, soya, and dark chocolate contain a lot of potassium. You can also sprinkle potassium over your food, just like salt – although potassium tastes a little more bitter than salt (you can buy potassium salt in most supermarkets). Keep in mind not to exceed the recommended daily dose of potassium. This recommended dose is about 3 to 4 g of potassium per day. But a gram-and-a-half can make a lot of difference. An extensive study proved that people who took 1.64 g of potassium more than they used to had 21 percent less chance of getting a stroke.[194]

SUMMARY

According to some recent studies, **the intake of salt** does not play such an important part in the prevention of cardiovascular diseases in healthy people.

In the West the intake of **salt** (sodium chloride) is too high, and the intake of potassium is too low.

Potassium deficiency is said to also play an important part in causing high blood pressure than only eating too much salt.

Products that contain a **lot of potassium** are fruits and vegetables such as apricots, pears, mangos, raisins, bananas, figs, tomatoes, avocados and soya.

A **good balance** between salt and potassium reduces the risk of getting a stroke.

Herbs

Herbs have always been the unsung heroes of the kitchen. At least, as far as our health is concerned. We think that herbs are great to enhance the taste of our food, but because we just use very small doses to season our food it seems as if herbs don't contribute a lot towards our health. However, this is a misconception. Herbs can have an important effect on the health of the body. Many herbs have powerful anti-cancer properties. This is also true of frequently used herbs such as parsley, thyme, rosemary, basil, oregano, marjoram or mint.

Take parsley. Parsley contains a substance with powerful anti-cancer properties, called *apigenin*. Apigenin, among other things, slows down the formation of blood vessels (*angiogenesis*) around tumours. As we have seen, tumours consist of runaway cells that won't stop dividing themselves; in the end, they become so big that they clamp off important nerves or arteries, as a result of which the patient dies. But cancer cells need nutrients and oxygen too. These are transported through the blood vessels. That is why cancer cells secrete substances that make sure the blood vessels surrounding them will grow properly, so the nutrients they need in order to grow faster can be supplied in sufficient quantities. Doctors call this kind of growth of the blood vessels angiogenesis. It's important to know that this growth is caused by *growth factors*, substances that will stimulate the growth of blood vessels.

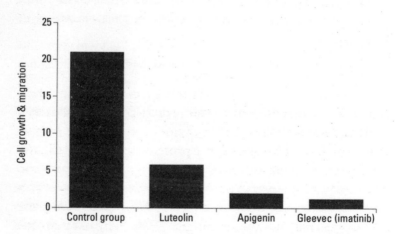

Apigenin is almost as effective in slowing down cell growth as Gleevec (imatinib). Luteolin is another substance that also inhibits cell growth, and is found in vegetables and herbs. The highest bar represents the cells to which no growth-inhibiting substances, such as apigenin, have been administered (the control group). *Source: The dietary flavones apigenin and luteolin impair smooth muscle cell migration and VEGF expression through inhibition of PDGFR-[beta] phosphorylation*, Cancer Prevention Research, *2008*

So, apigenin slows down these growth factors, and in this way the growth of the blood vessels is stemmed, and the growth of tumours is restrained too. It gets even better, since apigenin is almost as good at this as *imatinib*.

Imatinib is a famous cancer drug; basically, it's a molecule that has been designed in labs, through the use of computers, databases, and other types of high-tech methods, and it is part of the new generation of anti-cancer drugs. Imatinib works differently from the usual anti-cancer drugs, the *chemotherapeutics*. Chemotherapeutics are actually very toxic substances that poison the cells in the body. Cancer cells are much more sensitive to the toxic chemotherapeutics than body cells, because cancer cells grow so fast and divide in an uncontrolled way. But the problem is that chemotherapeutics are also toxic to the regular body cells and that is why patients become so sick when they are treated with these drugs.

However, the new types of anti-cancer medication, such as imatinib, deal with cancer in a totally different way: molecules like imatinib inhibit the proteins that regulate cell growth. Usually, these growth factors are a kind of protein that 'switch on' or trigger other proteins within the cell, and start them growing. With cancers, this growth is much too powerful and uncontrolled. You can think of imatinib as a growth factor inhibitor that deals with specific, rapidly growing cancer cells. An important reason why imatinib can fight tumour growth is because it also slows down the growth of the blood vessel cells around tumours.[195]

Doctors call imatinib a 'wonder drug'. The drug was portrayed on the cover of *Time Magazine* as 'the magic bullet' that could be used for a certain type of cancer (chronic myeloid leukaemia, CML). The team of top scientists that have developed imatinib received a prestigious award in 2009, 'for turning a fatal cancer into a manageable chronic disease'. It is therefore interesting that apigenin, a substance present in parsley, slows down the growth and migration of the muscle cells in the blood vessels too and is almost as powerful as imatinib (whose brand name is 'Gleevec') in doing this.

In the words of the researchers:

'Interestingly, the inhibitory effects of apigenin and luteolin on pulmonary aortic smooth muscle cell [a type of cells found in arteries] migration were similar to that achieved by Gleevec, a drug used clinically for the treatment of chronic myeloid leukemia.'[196]

What is remarkable is that to achieve this effect you don't need to use large concentrations of apigenin. The required concentration of apigenin in the blood to achieve the effects mentioned above can be attained by eating a regular dose of parsley, no more than the usual dose accompanying a meal. So it isn't necessary to directly inject a purified, high-grade apigenin concentrate into the bloodstream, for example. Additionally, studies show that apigenin does not only slow the growth of blood vessels around the tumours, but also the growth of tumours themselves.[197-198]

Naturally, this doesn't mean that CML patients just need to eat nothing but ground parsley leaves from now on. But this study, like many others, demonstrates that herbs contain a multitude of substances that can decrease the risk of getting cancer, and can restrain the growth of cancer cells. The pharmaceutical industry is well aware of this as well. Various companies are busy developing drugs that have a molecular structure that is very similar to the flavonoids that are present in herbs and vegetables. One of the drugs that are being developed is called *alvocidib*, also called *flavopiridol*. 'Flavo' refers to flavonoids. Alvocidib is a molecule that looks a lot like flavonoids such as apigenin. But the difference lies in the fact that the scientists have attached a small group of atoms to this molecule, in order to enhance the efficacy and absorption of the drug and to define it as an original, self-made molecule that can be patented. You don't need to be a biochemist to see that the backbone of alvocidib has a strong resemblance to that of apigenin and other flavonoids:

Apigenin (left) and to the right alvocidib, a new anti-cancer drug.

It's interesting that the pharma-industry is also studying the effects of alvocidib on diseases other than cancer. And these all happen to be ageing diseases as well, such as arterio sclerosis[199] (hardening of the arteries), and inflammatory diseases such as arthritis.[200] Inflammatory diseases and arteriosclerosis are not coincidentally diseases of which the risk can greatly be reduced by adhering to a healthy diet, consisting of vegetables, fruit and herbs that are rich in flavonoids.

But what is the exact reason for the fact that herbs can reduce the risk of getting cancer? It is often said that a substance protects one from cancer because it is an 'antioxidant'. The word 'antioxidant' has become very fashionable nowadays, and is used in all sorts of explanations. This is not a good thing, since it prevents people from thinking things through a bit further. The ageing process? Not enough antioxidants! Cancer? Too few antioxidants! Rheumatism, wrinkles or damaged DNA? Didn't get enough antioxidants! In the old days we invoked a deity to explain sickness and health. Nowadays we call upon antioxidants to provide an explanation for sickness, health, and old age.

But there is more to a substance that wards off cancer than it just being an antioxidant. For instance, such a substance can act upon the following mechanisms within a cell:

1 The substance concerned will slow down *kinases*. Kinases are proteins that 'switch on' all sorts of other proteins. As a result, kinases activate the cell, and cells can grow and divide.

2 The substance makes sure the *cell check proteins* do a better job. 'Cell check proteins' are proteins that regulate the growth cycle of cells. If this growth cycle is disrupted, cells can go rogue and cancer may develop.

3 The substance inhibits *inflammatory proteins*. Inflammatory proteins are proteins that activate or attract white blood cells. Cancer cells flourish best in an environment where there is a lot of inflammation. That is to say, inflammation on a molecular level, so no big festering wounds, but damaged cell walls and things like that, where white blood cells try to squeeze in to clean up the mess.

There is a substance that responds to all three of these mechanisms. This substance is called *curcumin*. Curcumin is a molecule derived from turmeric, a bright yellow powder that is made from the yellow roots of a plant that belongs

Curcumin

to the ginger family. This powder is used to produce curry sauces, among other things.

Curcumin slows down kinases,[201–202] regulates the cell check proteins[203], and curbs inflammatory proteins[204]. Curcumin does not just protect against cancer; it is also said to be able to slow down ageing diseases such as Alzheimer's. Studies have demonstrated that curcumin reduces the clumping of proteins in the brain that cause Alzheimer's. Alzheimer's disease originates from proteins that clump together in and around the brain cells, eventually suffocating these cells. These proteins are called *amyloid-beta* (Ab). Research on mice with Alzheimer's demonstrates curcumin reduces the deposits of amyloid-beta proteins by 43 percent.[205–206]

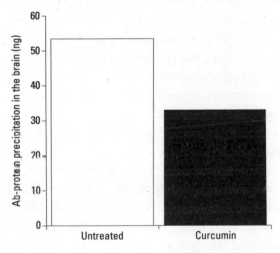

Curcumin reduces the formation of AB-plaques in mice, almost by half (black bar). *Source: The curry spice curcumin reduces oxidative damage and amyloid pathology in an Alzheimer transgenic mouse,* The Journal of Neuroscience, *2001*

Another study demonstrates that rats that are fed curcumin form less *lipofuscin*.[207] Lipofuscin is also called the 'ageing pigment'. It is a waste product that accumulates in our cells, as a result of the ageing process. It's a mishmash of protein clutter and oxidised fats that can occupy up to 70 percent of the cell volume in older people. In fact, lipofuscin is one of the causes of the ageing process. There are just a few substances that can slow down the formation of lipofuscin, but curcumin is said to have that quality.

In India, they didn't bide their time waiting for the results of these scientific studies. Indians have been eating turmeric for thousands of years, and this substance is an important ingredient in their dishes. Turmeric can be used in various dishes, in meat and fish alike, and in soups and vegetable dishes too.

Curcumin is a slightly fatty substance (which enables this substance to pass through the blood-brain barrier). In order to stimulate absorption by the gut, it is best to eat turmeric together with other fatty substances, such as olive oil. Black pepper also increases the absorption of curcumin; it ensures that curcumin is absorbed up to 20 times better.[208] You could season your salad with a teaspoon of turmeric every day, for instance, and add some black pepper and olive oil. Or you can add a teaspoon of turmeric to your vegetable soup or juice, with some black pepper too.

Curcumin is not the only substance that can affect Alzheimer's, and inflammation of the brain. Herbs such as thyme, camomile, peppermint, rosemary, and oregano contain numerous flavonoids, such as *luteolin*, that can reduce inflammation in the brain. For example, for four weeks luteolin was administered to older mice. Their working memory improved, and the inflammatory parameters in their brains fell to the levels found in young mice.[209] Some substances found in cinnamon can slow down Alzheimer's disease as well[210] , and also fight other ageing diseases, such as diabetes type-2.[211] A cinnamon extract caused 63 percent less clustering of amyloid-beta in mice with Alzheimer's, according to a study published in *PLoS Biology*.

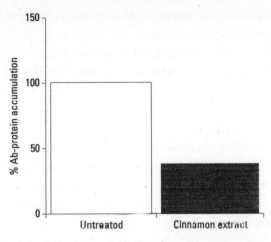

63 percent reduction of amyloid-beta clustering in mice with Alzheimer's that were fed a cinnamon extract. *Source: Orally administrated cinnamon extract reduces [Greek-beta]-amyloid oligomerisation and corrects cognitive impairment in Alzheimer's disease animal models*, PLoS Biology, 2011

In short, herbs are not merely useful as taste enhancers, but can also help to prevent numerous chronic and ageing diseases.

SUMMARY

Herbs, such as parsley, thyme, rosemary, turmeric, basil, oregano, cinnamon, marjoram or mint:
- can reduce the risk of getting **cancer** and slow down the growth of cancer cells;
- have a positive effect on **ageing diseases**, such as dementia or diabetes.

Curcumin, a substance derived from the yellow roots of a ginger-like plant:
- slows down the production of **inflammatory proteins**;

Continued overleaf

- slows down the production of proteins that enhance **cell and cancer growth**;
- reduces the clustering of **Alzheimer**-proteins;
- makes sure that less **lipofuscin** (an 'ageing pigment') piles up in the cells.

Garlic, onions, capers ...

Garlic and onions are full of toxic substances. Sheep, cows, and cats that eat garlic or onions can even develop hemolysis, where the rood blood cells burst because of the damage caused to them by the toxins in garlic and onions. So, one can assume that garlic isn't just toxic to vampires.

But taken in small quantities, garlic and onions are healthy, since they tickle and activate the defensive mechanisms of our cells (this principle has already been discussed quite often in this book).

Because garlic and onions are slightly toxic, they also activate certain enzymes in the liver that specialise in breaking down (other) toxic substances that can damage the body. Also, garlic and onions activate glutathione peroxidase, an important substance that breaks down free radicals.

People who eat at least half an onion a day, had 50 percent less chance of getting stomach cancer, according to a study published in *Gastroentorology*.[212] Onions also protect the body from cancers that go beyond the digestive tract, such as breast cancer.[213] Garlic is best known for its effect on the cardiovascular system. Garlic lowers the blood pressure and reduces the stickiness of the blood platelets, which has been proven in laboratories (in vitro), and with experimental subjects (in vivo).[214] So, garlic can be used to prevent cardiovascular diseases. Of course, garlic will not be able to undo the effects of someone who has already suffered a heart attack: the substances in garlic will not be able to work miracles and regenerate the dead heart muscle tissue. This is what Dr Khalid Rahman, a biochemist and researcher at Liverpool University, has to say about garlic:

'Garlic on its own will not treat cardiovascular disease but may prevent or delay its onset. However, evidence is emerging that it may be beneficial as an adjunctive treatment in the management of cardiovascular disease. Since garlic has so many other medicinal properties [...], regular consumption of garlic may reduce other chronic diseases associated with oxidative stress such as cancer and diabetes.' Dr Khalid Rahman (*Source: Vitasearch, The experts speak: cardiovascular disease and garlic*)

The substances in garlic are at their most powerful when the garlic is eaten raw, for example, as an addition to a salad, to legumes, or in healthy sauces and vinaigrettes. Adding garlic to oils, such as olive oil, promotes the absorption of the fat-soluble substances in garlic.

SUMMARY

Onions and garlic are slightly toxic, which causes them to:
- tickle the defence mechanisms of our cells, so that free radicals can be broken down more quickly;
- activate enzymes (proteins) that break down toxic substances in the liver.

This will result in reducing the chance of getting **cancer** and **chronic diseases**, such as cardiovascular diseases and diabetes.

Garlic especially reduces the risk of getting **cardiovascular diseases**, by inhibiting the aggregation of blood platelets, among other things.

Step 6: Dietary supplements

Medication

versus

'Smart' dietary supplements

Dietary supplements

Dietary supplements constitute an industry worth billions. And that's why we are greatly encouraged from all sides to improve our health by swallowing all sorts of pills. Do we really need all these dietary supplements? Yes and no. Sadly, this problem is mostly discussed in a black-and-white manner. On the white side you will find the anti-ageing and health gurus who are fierce supporters of dietary supplements and 'antioxidants', in the shape of wholesome vitamin E, grams of vitamin C, and the 'fortifying' co-enzyme $Q10$. On the opposite side we find the critics who throw out the good with the bad and claim that all dietary supplements are rubbish, and that a healthy, varied diet suffices.

Let me put it this way: the truth can be found somewhere in the middle. Some dietary supplements are advisable. These are the dietary supplements that have proved to be beneficial to our health (demonstrated by extensive scientific studies). Usually, these are also the substances that are clearly demonstrated to be deficient in large parts of the population, although this 'deficiency' needs to be interpreted in the right way. Indeed, there are two types of deficiencies: the 'gross deficiencies' and the 'suboptimal levels'. When doctors and governments refer to deficiencies, they usually mean gross deficiencies. These gross deficiencies are serious deficiencies, often so serious as to cause illness. On the other hand, a suboptimal level of some nutrients is a mild deficiency that doesn't make you ill in most cases, but that can weaken your health in the long run, and speed up the ageing process. These suboptimal deficiencies are much more common than the official deficiencies.

Let's take vitamin D, for example. Lots of government institutions and health care facilities define it as a deficiency if a person has less than 10 nanograms of vitamin D per millilitre (ng/ml) in his or her blood. This person is said to be vitamin D-deficient. This means that he or she has so little vitamin D that the person can contract rachitis (commonly known as rickets), a disease that causes deformation of the bones, since vitamin D is necessary to absorb calcium from the gut, among other things. According to a study carried out in a medium-sized town in the United Kingdom, one in four Brits has a vitamin D deficiency (that is to say, their blood contains less than 10 ng/ml of vitamin D).[215] And this is even measured at the end of the summer (vitamin D is produced through sunlight on the skin). Thus, those who are opposed to dietary supplements draw the conclusion that most Brits (specifically, 75 percent) do not need extra vitamin D. But there is a big difference between a deficiency and a suboptimal quantity of a certain substance in your blood. A study that was published in *The Lancet* demonstrated that people with less than 20 ng/ml vitamin D run twice as much risk of getting bowel cancer as those who have sufficient vitamin D.[216] Another study, also published in *The Lancet*, recommends an optimal dose of vitamin D in the blood of about 30 to 60 ng/ml.[217] Inhabitants of sunny countries (since people produce vitamin D through sunlight) even have an average dose of 50 to 90 ng/ml vitamin D in their blood. In a word, 'only' a quarter of the British population has an 'official' deficiency (less than 10 ng/ml in the blood), but a large majority of the English people is said to have a suboptimal dose of vitamin D in their blood (less than 30 to 60 ng/ml), which increases the risk of getting cancer (and various other chronic diseases, as we shall see). So, when the government and medical labs talk about a deficiency, they usually mean a serious, gross deficiency. A much larger percentage of the population may not be deficient, but does exhibit suboptimal levels that may lead to health problems in the longer run.

This was the main difference between suboptimal levels and deficiencies. But there are quite a lot of other misconceptions concerning dietary supplements. People often state that dietary supplements are not necessary if you stick to a healthy and varied diet. But this is hardly ever the case. First, almost no one eats really healthy and varied food, including those who think they adhere to a healthy and varied diet. If you really want to eat healthy and varied food, you would need to eat seaweed every week, in order to ingest sufficient iodine. On top of that a lot of dried shiitake mushrooms in winter, to ensure you have sufficient vitamin D in your blood. And also more than half a kilo of vegetables, for the necessary doses of magnesium and folic acid. Very few people eat seaweed every week, and neither do they eat dried mushrooms and heaps of vegetables. Furthermore, we could ask ourselves whether our healthy food is still that healthy nowadays. Fruits and vegetables contain much fewer vitamins and minerals than previously, since they are over-cultivated. For decades now, fruits and vegetables have been selected in such as way as to look nice, firm, and large on the shelves in the supermarket. They are no longer selected for their nutritional value. Also, modern cultivation techniques have lead to the growing of vegetables and fruit on over-cultivated agricultural soil, low in vitamins and minerals, which in turn causes vegetables and fruit to contain fewer vitamins and minerals, too.[218-219] For example, since 1940, vegetables contain 24 percent less magnesium,

27 percent less iron, 46 percent less calcium, and 76 percent less copper.

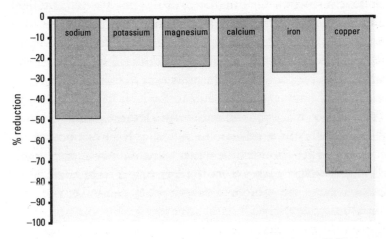

Vegetables contain many fewer minerals than in 1940. *Source: A study on the mineral depletion of the foods available to us as a nation over the period 1940 to 1991. David Thomas, based on studies by R.A. McCance and E.M. Widdowson, Ministry of Agriculture, Fisheries and Foods and the Royal Society of Chemistry*

We can conclude that today's tomato is no longer the same tomato as, say, 60 years ago. And that is why we can ask ourselves whether a 'healthy and varied diet' still contains sufficient nutrients. This is one of the reasons that the Harvard scientists have added a daily dose of multivitamin supplements to their food pyramid.

On the other hand, people run the risk of easing their conscience by using dietary supplements. Somebody who eats junk food might take a daily dose of multivitamins, in order to ingest 'sufficient' vitamins and minerals. Of course this is nonsense. During a well-known experiment, mice were fed an ostensibly 'optimal' diet: the diet contained all the minerals and vitamins that mice need, plus the necessary carbohydrates, calories and fats. Although the mice ingested all the required doses of vitamins and minerals, along with sufficient calories, the mice became ill after a while and died faster. Evidently, this occurred because a wholesome diet consists of many more substances than the couple of dozen familiar vitamins and minerals. Vegetables, fruits, nuts, cocoa and fatty fish contain a thousand types of flavonoids, fatty acids, and phytochemicals, most of which haven't been discovered yet. It is impossible to cram all these elements into a vitamin pill; you would need a pill of at least one pound.

There is yet another risk when using dietary supplements. And this danger comes from the fact that certain vitamins or antioxidants can inhibit the absorption of similar vitamins and antioxidants. Such as beta-carotene, for example. This substance is present in carrots, among other things, and gives them their orange colour. In the body, beta-carotene is transformed into vitamin A. Beta-carotene is sold as a dietary supplement because it is a 'wonderful antioxidant', and a precursor to vitamin A. According to a big vitamin website, beta-carotene is an 'anti-ageing micronutrient of the highest class, capable of fighting disease, improving the general health, and adding to a longer lifespan'.

But studies have demonstrated that people who take beta-carotene supplements run a greater risk of getting lung and prostate cancer and cardiovascular diseases, and exhibit larger numbers of deaths due to general causes.[220] A possible explanation for this is the existence of different types of carotene, while dietary supplements contain just one type of carotene, namely beta-carotene. As a result, the absorption proteins in the bowels and in our cells, which specialise in absorbing the carotene in our food, become satiated with this single type of beta-carotene, and are no longer capable of absorbing the other types of carotene. But these other carotenes are also essential in keeping the body healthy. It is believed that it isn't so much the beta-carotene, but rather the alpha-carotene that produces an anti-cancer effect. In short, a supplement that contains just a single type of carotene prevents the other carotenes from being absorbed, which increases the chances of getting cancer, or dying of general causes.

The same goes for vitamin E. Vitamin E supplements usually contain just a single type of vitamin E, which is alpha-tocopherol. But there are at least eight types of vitamin E known to us. If someone takes high daily doses of vitamin E, the receptors in the bowels that absorb the various types of vitamin E will become saturated with just one type of vitamin E. Because of this, your body will absorb less of the other types of vitamin E.

So, we have to deal with numerous, persistent myths regarding antioxidants and dietary supplements. Almost every health magazine or book will tell you that vitamin E is beneficial to the heart and blood vessels. Yes indeed, just the one type of vitamin E, that is to say alpha-tocopherol, could be capable of drastically reducing the risks to the cardiovascular system, according to a lot of health gurus. And there are actually several studies that prove that vitamin E can reduce the risk of getting cardiovascular disease. However, these studies are generally quite small and ill-conducted, and have been published in scientific journals that don't have a high impact. Let's just take a look at what the larger, extensive studies have to say. CHAOS (*Cambridge Heart Antioxidant Study*) demonstrated that vitamin E reduced the number of non-lethal heart attacks, but that vitamin E didn't have any effect on the number of *fatal* heart attacks. This is already quite strange in itself. Because if vitamin E was really beneficial to the cardiovascular system, you would expect the number of fatal heart attacks to decrease as well. Other studies do not demonstrate any effect at all for vitamin E. According to the GISSI study (11,324 test subjects), vitamin E doesn't protect people from heart attacks (although omega-3 fatty acids did have an effect). The same goes for HOPE (*Heart Outcomes Prevention Evaluation-study*), incorporating 9,541 test subjects who ingested vitamin E, and the MRC/BHF study (*Medical Research Council/British Heart Foundation-study*), that included 20,536 participants and where a mixture of vitamin E, vitamin C and beta-carotene was found to have no effect at all in preventing cardiovascular diseases. However, these days you can't find a health book or magazine article without reading about the beneficial effects of vitamin E for the heart and blood vessels, and then we haven't even yet considered the internet. On the other hand, half an hour's research in medical databases will tell you that vitamin E is not effective against cardiovascular disease, in spite of its great antioxidant activity.

Something similar has happened with vitamin C, the 'super-antioxidant'. Although the story is quite a bit more complicated here. Vitamin C is said to be so healthy because it is an antioxidant. The vitamin C supporters base their opinions on badly conducted, small-scale studies, and of course, on the books of Nobel Prize winner Linus Pauling too. Pauling is a chemist, and one of the very few people who has won two Nobel prizes, one for chemistry and one for peace. Pauling was a devout vitamin C user and wrote various health books that all became bestsellers. He recommended high doses of vitamin C, even multiple grams a day. Many people attribute the fact that he reached the wonderful age of 94 to the high doses of vitamin C he ingested. Although these people forgot to take into account that the average lifespan of a professor has always been ten years longer than that of a common labourer, and that Pauling lived a healthy life, and ate wholesome food.

But Pauling had a point, in a way. Indeed, he had demonstrated that cancer patients who were administered high doses of vitamin C, easily lived four times longer than patients who didn't get vitamin C. Although other studies that tried to confirm this finding, conducted in the famous Mayo Clinic in the USA, among other places, did not show any effect of vitamin C on cancer. At first glance, this would be the end of it. However, the big difference between these studies and Pauling's study was the way in which the vitamin C was administered; in the case of the Pauling study, the vitamin was injected intravenously, while the subjects in the other studies took the vitamin orally (through the mouth). Orally-administered doses seemed to have no effect, contrary to intravenous administration. Why would this be the case? Because vitamin C taken in high doses is in fact a pro-oxidant instead of an antioxidant! Yes indeed, vitamin C, the most famous antioxidant we know, can also be a pro-oxidant. When vitamin C is injected intravenously, the high doses will enter the bloodstream and vitamin C is capable of fighting cancer, because of its pro-oxidant activity. As we know, vitamin C can react with the iron within our cells. When vitamin C comes into contact with iron, it forms dangerous and highly reactive hydroxyl radicals (for the chemists among us: this happens through the Fenton's reaction). These hydroxyl radicals are free radicals that react with the DNA and the proteins in our cells, and in this way they damage them.

Cancer cells are especially sensitive to the pro-oxidant activity of vitamin C. This is due to the fast and uncontrolled growth of cancer cells, which prevents them from managing their iron levels properly. In cancer cells, the iron is no longer neatly packaged but it swings about in every direction, and so it can react with the vitamin C, in order to form free radicals that are extremely toxic to cancer cells. In short, vitamin C taken intravenously in high doses is just as toxic as a chemotherapeutic drug. Note that our regular cells can also be damaged by high doses of vitamin C, although less severely than cancer cells.

Nevertheless, a lot of health gurus were content to read only Pauling's study in which he described how vitamin C promoted the chances of surviving cancer. And because they are not knowledgeable with respect to the underlying biochemical mechanisms, they still claim that high doses of vitamin C are beneficial to your health.

There is yet another reason for not recommending high doses of vitamin C, since a surplus of vitamin C just leaves the body when you urinate. This is because the body strictly regulates the vitamin C levels (perhaps because the substance can also act as a toxic pro-oxidant). But of course this doesn't mean that we should throw away the good with the bad. On the contrary, in most countries the recommended daily dose of vitamin C is about 60mg. Which in fact is too low. So how have the government institutions computed this far too low dose? They based their recommendations on studies that demonstrate that a few dozen milligrams of vitamin C suffice to prevent scurvy, and that a person starts to secrete vitamin C when doses exceed 60mg. But this reasoning is faulty. Scurvy is a serious disease that develops when people have far too little vitamin C in their system. The gums start to bleed, internal bleeding in the stomach occurs, patients feel tired, weak and nauseous. In a word, this 60mg dose might be the minimal daily amount you need if you are an eighteenth-century sailor in the British navy, and have been floating in the Pacific for months. But for a person living in the twenty-first century who strives for optimum health, this dose is far too low. The body doesn't just need vitamin C to prevent scurvy, but also to produce neurotransmitters, for collagen synthesis, for energy metabolism, etc.

We could say that this recommended daily dose of 60mg, is far too small, whereas the grams of vitamin C prescribed by the health gurus are excessive. And in this way the general public is pulled back and forth in what looks like a serious debate, while the answer is actually quite simple.

The best thing to do is to take a medium dose of vitamin C, somewhere between 200 and 400mg per day. This is because studies have demonstrated that the body is completely satiated with a dose of 400mg of vitamin C (and not with a mere 60mg), and because vitamin C acts as a pro-oxidant instead of an antioxidant when the dose exceeds 500mg.[221] This daily dose can be doubled for people who smoke.

And this brings us to an important issue: using a surplus of most of the dietary supplements is primarily beneficial to the wallets of the supplements' manufacturers, whereas a lack of certain vitamins and minerals can be bad for your health, in the long term. That is why a multivitamin supplement is ideal, to make up for possible deficiencies of certain vitamins and minerals. Because these deficiencies surely exist, especially with the elderly. Older people don't absorb vitamins, minerals and phytochemicals as well, because the stomach produces less gastric acid and the bowels function less well. This reduced absorption already starts at about forty years of age.

In the table below we are the percentage of the population that ingests less than the EAR (estimated average requirements) of various vitamins and minerals. According to this table, half of the American population doesn't get enough magnesium. Almost every American has too little vitamin E, and half of the over-seventies have too little vitamin B6. Of course, the American way of living differs from the European lifestyle, but this is a trend towards which we all evolve.

In this respect, we should not forget that the estimated average requirements (EAR) are even lower than the recommended daily dose (RDD) recommended by the government. And, as we have seen way too often, the RDD is already much too low for many products, since it is primarily directed at deficiencies, and not at suboptimal shortages. If we were to survey the suboptimal shortages, the numbers in the table below would be even more alarming.

Lack of nutrients in the USA.

Nutrient	Population group	Percentage of population that ingests less than the *estimated average requirements*
Iron	Women 14-50 yrs	16
Magnesium	Everyone	56
Zinc	Everyone	12
Vitamin B6	Women older than 71 yrs	49
Vitamin C	Everyone	31
Vitamin B9 (folic acid)	Adult women	16
Vitamin E	Everyone	93

Source: Usual nutrient intakes from food compared to dietary reference intakes, The National Health and Nutrition Examination Survey, 2001–2002

Until now I have mainly discussed dietary supplements, especially the antioxidants. But there are various, more effective dietary supplements to be found. These are mainly substances of which a lot of people have suboptimal levels, or substances that affect the metabolism and in this way slow down the development of ageing diseases. As opposed to all these 'hyped' antioxidants I call these the 'smart dietary supplements'. Magnesium is an example of such a smart dietary supplement.

Magnesium has an important function in human metabolism. It stabilises adenosine triphosphate (ATP). ATP is also called 'the second most important molecule in the body, apart from DNA' by scientists. ATP is energy. Literally. ATP keeps almost all molecular reactions going. We eat food and breathe in oxygen in order to produce ATP. Magnesium sticks to the ATP molecules and stabilises them, so they can do a better job and the energy metabolism works at its best. Since magnesium is so important for the metabolism, it doesn't come as a surprise that a (suboptimal) lack of magnesium increases the risk of contracting cardiovascular diseases, type-2 diabetes, cancer, hypertension, osteoporosis, strokes, and other ageing diseases. For instance, people whose blood contains sufficient magnesium have 40 percent less chance of getting cardiovascular diseases, and 50 percent less chance of getting cancer.[222] Scientists recommend that we ingest at least 300 to 600mg magnesium daily. However, most dietary supplements don't contain such a dose because it will make them too large. And a lot of dietary supplements contain magnesium oxide, which is badly absorbed and can irritate the stomach and bowels (in medicine, magnesium oxide is even used as a laxative). Magnesium citrate doesn't bother the stomach and bowels as much, and is also better absorbed.

Another group of substances that play an important part in the metabolism are the B-vitamins. Vitamins B1, B2, B3 and B5 are the oil that grease the complex mechanism that is metabolism. Vitamins B9 and B12 are more involved in the production and integrity of the DNA. Shortages of B-vitamins mainly occur in tissue that is very metabolically active, such as the brain, the kidneys, or the heart. For example, a lack of vitamin B1 causes the Wernicke and Korsakov syndromes. This irreversible disease causes brain cells to die and damages the memory. Wernicke-Korsakov is often found in alcoholics, because they don't eat well (and mainly incur vitamin B1 deficiencies as a result of this). In the well-known Nurses' Health Study, women with the highest level of vitamin B9 had 45 percent less chance of getting a heart attack.[223] Similar percentages apply to vitamin B6 as well. According to a different study, published in *Neurology*, older people with low vitamin B12 levels suddenly had a six times greater chance of suffering from brain shrinkage.[224] Researchers from Oxford University who gave elderly people vitamin B, saw in the following years that these people's loss of brain volume was seven times smaller.[225] This is what professor Walter Willett from Harvard University, one of the main authorities in the field of nutrition and health, has to say about B-vitamins: 'Research increasingly proves that variousingredients of a regular multivitamin supplement – especially the vitamins B6, B12, B9, and vitamin D [see further on] – are essential factors in preventing heart diseases, cancer, osteoporosis, and other chronic diseases.'[98]

Since the B-vitamins are soluble in water, a surplus will always be secreted through the urine. This way, an overdose of B-vitamins hardly ever occurs. However, some very high doses may cause side effects, for example, more than 50 mg vitamin B6 per day (the recommended daily dose of B6 is 2 mg). The best thing to do is use a vitamin B-complex. This is a supplement that contains different B-vitamins. This is important, because the various B-vitamins enhance each other and cooperate to make the metabolic processes come along smoothly. There is a study that shows that dietary supplements containing B-vitamins do *not* diminish the risk of getting a heart attack.[226] This is quite logical, since in this study only vitamin B9 and B12 is given, not a vitamin B-complex which contains B1, B2, B3, B5, B6, B9 and B12. However, another study demonstrates that administering vitamin B9, B12 and at least one extra B-vitamin, namely B6, was associated with a 50 percent reduction in dying of a heart attack or cardiovascular disease.[227]

Iodine too, plays an important part in the human metabolism. It regulates the speed of the metabolism, and the body temperature. Because iodine is such an important factor in activating our energy processes, even a mild shortage (far from an official 'deficiency') can cause health problems, such as fatigue and concentration problems.[228] Despite the fact that iodine is an important substance, almost 70 percent of various European populations have an iodine shortage. The World Health Organisation (WHO) also acknowledges the fact that the iodine shortage in Europe is a big medical problem,[229] in spite of the addition of iodine to bread and salt in many countries. Although you need to be careful when you use iodine supplements. Too much iodine can be toxic, especially to the thyroid gland. 200 micrograms is the maximum dose for a dietary supplement.

Selenium is a mineral, just like iodine. Selenium is an interesting substance because it plays an important part in the *body's own* antioxidant system. Our body contains an extensive internal antioxidant system, in the form of various types of proteins that clear away free radicals. These are proteins such as *superoxide dismutases*, *catalases* and *glutathione peroxidases*, which specialise in intercepting free radicals and rendering them harmless. One of the reasons why most of the antioxidants from the supermarket don't work well and don't have a lot of impact on lifespan or the ageing process, is because they are nothing compared to the body's own antioxidant system that is already embedded in our cells. Often, the ingestion of antioxidants through supplements even weakens this internal antioxidant system, which in general renders us more vulnerable to damage done by free radicals. We have already discussed this principle at length.

One of the main proteins in the body's own antioxidant system is glutathione peroxidase. However, glutathione peroxidase requires selenium in order to function properly. By ingesting selenium the body's own antioxidant system is strengthened and this also means that free radicals can be cleared away much better.

Besides, selenium is very important to the immune system. We know that many proteins in the immune system need selenium in order to do their job properly. In this way, selenium prevents viruses from replicating and mutating, such as the coxsackie virus that can cause heart problems, the hepatitis B and C viruses that can cause liver failure and liver cancer, and the HIV virus that causes AIDS. Patients who have HIV and also a shortage of selenium, have a greater chance of dying, about twenty times(!) greater than HIV patients who do have sufficient selenium in their blood (of course, other factors have been taken into account, such as malnourishment).[230] According to another study, comprising 18,709 participants, the people with the highest concentration of selenium in their blood had a six times smaller chance of contracting a specific form of arthritis. Arthritis is an inflammatory disease, caused by the immune system attacking the joints.[231]

As we have stated lots of times in this book, the immune system is also an important factor in the prevention of cancer. This is because the immune system is continuously clearing away recently developed cancer cells, and prevents them from developing into a tumour. Studies demonstrate that selenium can offer protection against cancer. According to a Harvard study with 34,000 test subjects, those who had a lot of selenium in their blood had three times less chance of getting prostate cancer.[232] According to the *Nutritional Prevention of Cancer Trial* (the first really decent, 'double-blind, randomised, placebo-controlled intervention study' of the use of selenium in a western population), people who were administered selenium yeast supplements had 50 percent less chance of dying of cancer, 63 percent less chance of getting prostate cancer, 58 percent less chance of getting bowel cancer, and 46 percent less chance of getting lung cancer.[233]

However, according to the Selenium and Vitamin E Cancer Prevention Trial (SELECT), selenium did not reduce the risk of prostate cancer (and vitamin E even increased the risk of prostate cancer). Yet in this study another kind of selenium supplement was used, namely selenomethionine instead of selenium yeast. Because vitamin E increased the risk of prostate cancer, researchers also wonder why the combination of vitamin E and selenium didn't increase the risk of prostate cancer.

All in all, most Europeans don't use sufficient selenium on a daily basis. The main reason for this is that European soil contains little selenium.[234] Selenium is found in vegetables and grain, among other things, and the concentration in these plants depends on the amount of selenium in the soil. Usually, a person should ingest at least 75 micrograms of selenium every day. With this dose, the glutathione peroxidase (the body's own antioxidant protein, containing selenium) is saturated the most. But most Europeans don't ingest this high a dose:

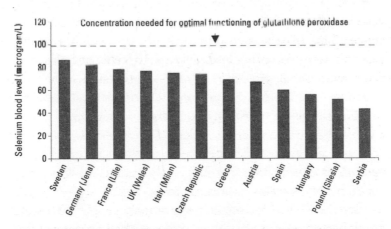

In many countries, the population ingests too little selenium to promote an optimal functioning of the body's own antioxidant proteins. *Source: The importance of selenium to human health*, The Lancet, 2000

Although there are some scientists who think this dose of 75 micrograms is not yet sufficient. In order to saturate the glutathione peroxidase in the cells as well (and not only the glutathione peroxidase in the blood), a daily dose of 100 micrograms should be taken.[235] And the anti-cancer and immune-stimulating effect of selenium only occurs with much higher doses, such as 200 micrograms a day. That is why some researchers recommend taking a total of 200 micrograms of selenium per day.

Recently, the European Union has issued a new guideline regarding the recommended dose of selenium. This has been established at 55 micrograms a day, which is even less than the 75 micrograms necessary to saturate the glutathione per-oxidise in the blood. Within ten years or so, this recommended dose is expected to be raised once again.

Selenium is a powerful substance, which also means that it can become toxic very fast. It is important not to overdose when taking selenium supplements. The maximum dose has been fixed at 400 micrograms a day, but it is best to restrict the use of selenium to 200 micrograms a day. This means that a dietary supplement can contain no more than 100 micrograms of selenium at most, since the daily meals and any other multivitamin supplements that are used also contain selenium. If you take dietary supplements, especially selenium and iodine, it is a good idea to consult a doctor who specialises in nutrition and preventive medicine.

Vitamin D is another interesting substance that has an important effect on the body. And a substance of which lots of people have a 'suboptimal shortage'.

Vitamin D is a fat-soluble vitamin that has a structure similar to that of a hormone. Vitamin D is produce in the skin, through sunlight. In the skin, pre-vitamin D molecules reside that are converted into vitamin D through sunlight (after having passed through the liver and kidneys). Vitamin

D is mainly known as the vitamin that is required to prevent rachitis. Rachitis is a disease that causes skewed growth of the bones, which often occurred among poor labourers' children in the nineteenth century, who didn't get enough sunlight because they had to work in the factory all day long. But in the last few decades, scientists have discovered that vitamin D does not only play a part in the development of bones and the calcium absorption. This vitamin also affects the immune system in a major way, and is of importance in relation to cancer, cardiovascular diseases and diabetes.

For instance, women who ingest sufficient vitamin D have a lower chance of contracting breast cancer (and a lower death rate).[236-237] Another study, published in *The Lancet*, demonstrated that babies that were given sufficient vitamin D supplements had 80 percent (!) less chance of contracting type-1 diabetes.

These studies are interesting with respect to a certain statement I often hear, that dietary supplements are 'unnatural'. In other words, if we would let nature take its course, (foremost, by eating and living healthily), dietary supplements would not be necessary. I can think of two arguments that counter this statement. First, most people no longer live in such a 'natural' way. It's no coincidence that the vitamin D babies in *The Lancet* diabetes study all live in Scandinavia, where there is less sunlight. The human race originates from Africa, where more than enough sunlight was available. Over the course of several millennia, man pushed his boundaries, literally and figuratively: African humans started to colonise Europe about 65,000 years ago. Because Europe was an 'unnatural' environment for us (with much less sunlight), we have developed white skin, in order to absorb more vitamin D. You could say that vitamin D is the reason why white people exist. But nowadays we lead even (much) more unnatural lives, inside our houses and sheltered by windows, devoid of sunlight. When we go outside for a second, it's just to get in the car or catch a train. And I haven't even started complaining about the English weather yet. In a word, our lifestyle doesn't remotely resemble that of our ancestors, especially if we take into account that we eat unhealthy food, don't get enough exercise, and experience a lot of new types of stress. The thesis that we 'should let nature take its course' no longer applies, simply because we no longer live in a natural way. So we could use some 'unnatural intervention' in the form of dietary supplements, just like we could go to the gym more, eat steamed foods, or install a wake-up light on our nightstand.

Secondly, we shouldn't forget that nature does not always have our best interests at heart. Mother Nature has only one issue on her mind and that is the procreation of the individual members of the species (nature doesn't even care a lot about the 'species as a whole', which is an important nuance). That is why we look beautiful and young up to the age of thirty. After that, the decline starts, and the ageing process too, since Mother Nature can no longer be bothered with us: she expects us to have reproduced ourselves by that time. In short, from thirty years onward, everything starts going down: waste materials accumulate in our cells, our nervous reflexes deteriorate, muscle mass is increasingly replaced by fat, the skin loses its elasticity, etc. Holding on to the statement that 'we should let nature take its course', without trying to be healthy and have a healthy metabolism as best we can, would be a pity since we just have this one life and one body, and since nature couldn't care less once we are past thirty years of age.

But let's get back to vitamin D. How can vitamin D actually reduce the risk of getting cancer and type-1 diabetes? The immune system plays a major part in both diseases. Vitamin D regulates, calibrates and reinforces the immune system. The immune system clears away sources of young cancer cells, which explains the effect of vitamin D on cancer.

With type-1 diabetes, on the other hand, the immune system attacks the body's own *beta cells* in the pancreas. These beta cells produce insulin. If these beta cells are destroyed by the immune system, patients are not capable of producing insulin, which allows the sugar to remain floating in the bloodstream and cause damage all over the place.

Vitamin D also plays an important part in other immune system diseases, such as multiple sclerosis (MS). With MS, the nerves are attacked by the immune system, which leads to nerve damage all over the body. This causes people who suffer from MS to suddenly get double vision, have paralysis in a limb, get problems with balance, etc. Studies have demonstrated that people with vitamin D levels higher than 40 ng/ml have 62 percent less chance of contracting MS, compared to people who have lower levels of vitamin D.[238] And indeed, a recent study published in *Neurology* proved that people who suffer from MS and are administered vitamin D, suffer 41 percent less from MS attacks.[239] That is a better result than that of the 'regular' treatment of MS, where the drug called beta interferon is administered. Beta interferon reduces the MS attacks up to a maximum of 30 percent. Moreover, this drug has a lot of side effects (80 percent of the patients feel 'like they have the flu', are fatigued and have muscle aches), and it costs a huge amount of money (852 euro for four doses). Vitamin D capsules cost less than one euro a piece and have barely any side effects.

Vitamin D also plays a part in other diseases that involve the immune system. For instance, a too low dose of vitamin D can have influence on depression, especially with older people. That is because the elderly don't get outside in the sunlight often enough, and they don't transform vitamin D in their skin as well as younger people do. Some medical journals even recommend to perform a routine check of vitamin D levels in each elderly person with depression.[240] Vitamin D is also involved in cardiovascular diseases. People who have sufficient vitamin D in their blood have more than 50 percent less chance of getting a heart attack.[241]

In short, Dr John White and Dr Luz Tavera-Mendoza, two vitamin D researchers, say this:

A clear relationship between low levels of vitamin D and cancers, auto-immunity, infectious diseases and other types of illnesses suggests that the currently recommended daily dose of this essential nutritional substance needs to be revised. (*Source: Scientific American*, November 2007)

And indeed, as is the case with many other nutritional substances, the recommended daily dose is much too low. That is why vitamin researchers recommend taking vitamin D in supplements with a higher dose than the government recommends. The daily dose recommended in many countries is 400 units (or 10 micrograms) of vitamin D a day. But several vitamin D experts recommend taking 2,000 units a day,[238] that is a lot more.

Although we have to say that a surplus of vitamin D can also be dangerous, which goes for many other substances too. Vitamin D is a fat-soluble vitamin. This means that a surplus is not secreted along with the urine (as is the case with water soluble B-vitamins or vitamin C), but is stored in the fat and the liver. But to incur an overdose of vitamin D you would need to ingest a very large amount of this vitamin, more than 10,000 units a day, over many months.[242] The toxic dose in the blood will then be higher than 150 ng/ml. The ideal level of vitamin D in the blood lies between 30 and 60 ng/ml. We have seen that in sunny countries, the amount of vitamin D even lies between 50 and 90 ng/ml. This fact could be an explanation for the so-called 'French paradox', that is to say, the observation that cardiovascular diseases are less common in the sunny countries surrounding the Mediterranean Sea. After all, we have seen that vitamin D also protects against cardiovascular diseases. This is the reason why some scientists believe that the lower occurrence of heart attacks among the Mediterranean population is more due to their exposure to sunlight than to their Mediterranean diet of pasta and olive oil. And indeed, in countries that don't have as much sunlight, such as Scandinavia or Canada, a larger amount of diseases are found where vitamin D plays a part. Multiple sclerosis for instance, occurs thirteen times less in Brazil than in Canada.[243]

Sometimes people claim that it isn't necessary to take vitamin D supplements, because lying in the sun for a short while is enough to produce a large dose of vitamin D. Indeed, 20 minutes in the afternoon sun can cause the body to produce 10,000 units of vitamin D. But only on the condition that you lie in the sun completely naked, and that it's equatorial sunlight. Because many countries are situated on a higher degree of latitude, the sunlight penetrates the earth at a more slanting angle, which makes it less powerful. Moreover, if the sun shines at all, it's usually not the entire body that is exposed to the sun. As a result, many people produce too little vitamin D, even during the summer. Which explains why such a large part of the population has a suboptimal shortage of vitamin D.

As far as dietary supplements are concerned, we can conclude that studies have shown that the recommendations issued by the government as to the various doses, are often too low. On the other hand, we are flooded with antioxidant supplements that are said to protect us against cardiovascular disease, cognitive deterioration and ageing. Most of these assertions are untrue or highly exaggerated.

It is much more important to prevent suboptimal shortages (and not just serious deficiencies) of certain substances. These substances are the B-vitamins, magnesium, iodine, selenium and vitamin D, rather than the popular antioxidants, such as vitamin C, vitamin A or E. Many of these substances play an important part in the metabolisms and the body's own antioxidant system. Moreover, suboptimal shortages of these substances are responsible for our health slowly deteriorating over the decades. As a result, our ageing process will speed up, or we might become more susceptible to numerous chronic diseases.

SUMMARY

Regarding dietary supplements, studies published in reputed journals have demonstrated that:
- antioxidants do not prolong the lifespan;
- most antioxidants are unhealthy in the long term;
- vitamin E does not play a part in the prevention of cardiovascular diseases;
- vitamin C can act as a pro-oxidant (especially in high doses).

It is important to distinguish between these two terms:
- **deficiencies**: these are major shortages that lead to health problems in the short term;
- **suboptimal shortages**: these are minor shortages that lead to health problems in the long term, and that can speed up the ageing process.

Smart dietary supplements:
- specifically target the **metabolism** (such as magnesium, B-vitamins and iodine);
- activate the body's own **antioxidant system** (such as selenium);
- replenish frequently occurring **suboptimal shortages** (such as a shortage of vitamin D).

Note that none of these smart dietary supplements are 'antioxidants'.

Examples of smart dietary supplements are:
- **magnesium** (300 to 600 mg a day, as magnesium citrate);
- a **B-vitamin complex** containing vitamins B1, B2, B3, B5, B6, B9 and B12 (the recommended daily dose multiplied by a few times);
- **iodine** (up to a maximum of 200 micrograms per day);
- **selenium** (up to a maximum of 100 micrograms per day);
- **vitamin D** (2000 units or 50 micrograms per day);
- completed by an optional **multivitamin complex** in order to ingest most necessary vitamins and minerals.

Vegetables, fruit, nuts, mushrooms, fatty fish and legumes contain thousands of other substances that are not found in dietary supplements, and that have an important effect on our health.

Medication

We use far too many drugs. One of the reasons is that we think this medication will cure us. But this is a huge misconception. Most drugs don't cure diseases, but suppress the illnesses' symptoms. And we also tend to underestimate the side effects of medication. This is because our body is very strong. On a molecular level, a lot of damage needs to have been done before we start feeling ill and start feeling the side effects of certain drugs. But this doesn't alter the fact that these side effects can be harmful to our health in the long term. Medication will always be a product that is foreign to the body and disrupts a delicate balance.

Let's take a look at hypertension medication, for instance. A hypertension drug does not 'cure' high blood pressure. A hypertension drug such as the 'calcium channel blockers, is a pill that contains trillions of a certain type of molecule. When you take that pill, these molecules are absorbed in the bloodstream and they will stick to certain proteins. These proteins are small calcium channels in the muscle cells of your blood vessels. The calcium channels will be blocked due to this drug, and less calcium will enter the muscle cells, which causes the muscle cells in the blood vessels to relax, the blood vessels to expand, and the blood pressure to drop. In a word, hypertension medication lowers your blood pressure by disrupting a certain mechanism on a molecular level, and not by tackling the cause of the hypertension. Furthermore, these anti-hypertension molecules not only attach themselves to the calcium channels in the muscle cells of the blood vessels, but also to those in the heart (which causes the heart to contract in a less powerful way), the brain (which changes the transmission of brain signals), the bowels (which causes constipation), etc. This causes numerous side effects, while this medication does not solve the heart of the problem, that is to say, the cause of the hypertension. The causes are mainly found in unhealthy eating habits.

This applies to almost all types of medication, except for antibiotics, or chemotherapeutics, that are actually capable of curing some types of diseases. But most drugs don't address the cause, they mainly suppress the symptoms. Gastric acid inhibitors suppress the production of gastric acid and don't treat the cause of too much (or too little?) gastric acid. Sleeping pills stun the patients, but don't solve the sleeping problems. Cholesterol inhibitors slow down the production of cholesterol, but don't suppress the unhealthy eating habits that lead to cardiovascular disease.

Apart from the fact that most medication just suppresses symptoms, drugs can have lots of side effects that we often underestimate. I will discuss some examples of frequently used drugs that are prescribed exactly because they have so few side effects, such as the *proton pump inhibitors* (PPIs), which are bestselling drugs.

The proton pump inhibitors are a class of gastric acid inhibitors. The molecules of which these gastric acid inhibitors consist stick to a certain type of proteins in the wall of our stomach cells. These proteins produce gastric acid.

The PPI-molecules prevent these proteins from doing their job, so less gastric acid is produced and patients suffer less from acid reflux. Sounds great, especially because the PPI-molecules mainly target the proteins in the stomach cells, and in this way cause few side effects.

But the problem is that the body needs gastric acid. For example, gastric acid takes care of the absorption of minerals, such as calcium, magnesium and iron. These minerals react with the gastric acid which changes their charge. It is due to this altered charge that minerals can be absorbed by the bowels. If a person produces less gastric acid, as a consequence he also absorbs fewer minerals, minerals such as calcium, for example, and this leads to a higher risk of bone fractures. A study that was published in 2006 in the reputed *Journal of the American Medical Association* demonstrated that people who use proton pump inhibitors have almost three times as much chance of getting hip fractures.[244] Apart from absorbing minerals, gastric acid also kills the bacteria we absorb through our food. You could say that the gastric acid 'sterilises' the food. But if less gastric acid is produced because of the ingestion of PPIs, a larger amount of bacteria can pass through the stomach without being attacked, and colonise the gut. From the gut, these bacteria can also leak into the bloodstream and cause infections.[245] Or they can be coughed up from the stomach and end up in the lungs, causing pneumonia. It has been shown that people who take PPIs have two to five times more chance of contracting pneumonia (depending on their age and the duration of the treatment)[246] and three times more chance of getting an infection caused by a dangerous gut bacterium (*Clostridium difficile*).[247] Less dangerous, but not very good for your health, is the colonisation of the bowels by unhealthy bacteria. According to a study, people who took PPIs had eight times more chance of getting an infection of the small intestines because of an overgrowth of intestinal bacteria.

Such an infection can cause severe diarrhoea, but can also go by unnoticed, without any symptoms, and can reduce the absorption of important vitamins and minerals.

And we would almost forget it, but we also need gastric acid to digest our food. By ingesting PPIs, our food is insufficiently digested, and too large chunks of badly digested proteins will end up in the large intestines. This way, protein fragments not only cause bacterial overgrowth, but they cause the immune system to produce antibodies too, to fight these protein fragments. These antibodies are also capable of attacking similar proteins in the body, which can cause auto-immune diseases, such as asthma, coeliac disease (gluten sensitivity) or arthritis.

So, in this way PPIs can enhance the risk of getting fractures, infections and auto-immune diseases. Nevertheless, gastric acid inhibitors are known as drugs that have very few side effects. And that is true. Compared to a lot of other drugs, PPIs are quite harmless and can be taken for many months or years running. But note that medication always disrupts important balances within the body. It is particularly the long-term effects of medication that are ignored, because none of the scientific studies has a duration that is long enough to discover that reduced absorption of magnesium, due to using PPIs, may cause a bigger risk of getting a cardio-vascular disease thirty years later on. Or to prove that the reduced absorption of calcium causes osteoporosis within twenty years. Moreover, these studies would cost loads of money because they have such a long duration and need to use a very large number of test subjects, since they can suffer from numerous other diseases and ailments over such a long period of time.

Another example of medication that is often prescribed are sleeping tablets. Sleep medication is one of the most frequently sold drugs, just like proton pump inhibitors. Sleeping pills are benzodiazepines, or substances derived from these. Benzodiazepines are molecules that stick to protein channels in the walls of our brain cells. This causes the channels to open up more, and in this way more negatively charged chlorine atoms can flow into the brain cells. These chlorine atoms will subsequently stick to proteins, which will diminish the activity of these proteins in the brain cells. This will render the brain cells less active, which makes us sleepy. So this medication doesn't address the sleeping problem itself, it just stuns the patient.

Sleeping pills cause people to fall asleep faster, but their sleep will be less deep and less refreshing. That is because these drugs alter the architecture of sleep. The sleep architecture indicates the various stages of sleep. Each sleeping phase has a specific duration and a characteristic EEG pattern (an EEG pattern is the result of the electrical activity of the brain). Benzodiazepines, for example, shorten the deep sleep (also called the delta wave sleep), which causes the patient to sleep less well.

Apart from this, sleeping pills are highly addictive. This means that within a week or so our brain cannot do without these pills. If we stop taking sleeping pills, the brain cells will even become overactive, which will make it even more difficult to fall asleep. And, to add insult to injury, in the longer term benzodiazepines cause forgetfulness, concentration disorders, drowsiness, dizziness (with a higher risk of falling, a big problem with the elderly), and paradoxically, also aggression, impulsive behaviour, and irritation. Many official bodies recommend that doctors not prescribe sleeping medication for longer than a week at a time. But I've met quite a lot of patients who use sleeping pills for years on end, and of course can no longer do without them. This disrupted sleep pattern can cause other types of problems in the long term, such as a higher risk of experiencing depression and fibromyalgy-like complaints.

By the way, withdrawal symptoms also occur with proton pump inhibitors. When patients stop taking PPIs, the 'brakes' on gastric acid production are taken off, and the suppression of the production of gastric acid (which can have gone on for years) stops. As a result, the patient suddenly produces too much gastric acid which forces him or her to start using the gastric acid inhibitors again.[224] At the same time, this is also the fascinating thing about the human body: you can hardly ever outsmart it. By taking medication, certain problems can be obscured, but in the long term, a whole new set of problems arises.

Let me just discuss one last type of medication that is also quite popular, namely the analgesic drug called paracetamol. Paracetamol is prescribed a lot because it is a very mild painkiller that has a lot fewer side effects than other painkillers, such as aspirin, diclofenac, ibuprofen or naproxen. Paracetamol has the most favourable side effects profile of all painkillers. Paracetamol is on the market in pills of 500 milligrams, and 1 gram (1000 mg). People often take the 1 gram tablets because their effect is stronger. They also take several pills a day, as the instructions state. However, paracetamol is harmful to the liver. Eight grams of paracetamol will suffice to cause acute liver failure. This means that a mere eight pills of 1 gram of paracetamol suffice to poison the liver in such a way that the liver cells die en masse, which can be lethal. In combination with alcohol, this effect is even more powerful. If somebody drinks alcohol, four tablets of 1 gram of paracetamol can suffice to cause acute liver failure.[248] In the US, paracetamol is also used as a suicide drug, because it is so toxic (and causes a very painful death). 10 to 15 grams of paracetamol is a lethal dose. And this is what the most guileless painkiller of them all can do.

Of course, ingesting a single tablet of paracetamol will not cause liver failure. And neither will it do any harm if four tablets are taken throughout a whole day. But it isn't hard to imagine that a single tablet is not particularly good for the liver either. Liver failure expresses this on a major scale, because the entire liver stops functioning. This means that the damage on a molecular level is so large that the liver cells dies en masse. The same type of molecular damage, but on a much smaller scale, starts to appear when you take one tablet of paracetamol. Fortunately, the human body is very resilient, so most people will not feel their liver suffer when they take some paracetamol. This is something you quickly learn as a doctor: the body can take on an awful lot, and patients can consider themselves lucky that they are usually ignorant of what happens on a molecular level when they take a certain medication (or suffer from some disease).

In short, we should stop taking medication without thinking it over thoroughly. As we have seen, even the most innocuous type of drug has its side effects. And the vast majority of drugs only suppress the symptoms, instead of tackling the causes. One of the big problems in our health-care system is the over-medication of patients. This surplus of medication often causes new health problems for the patient, which in turn leads to new prescriptions for new medication. Patients have to swallow pills that don't really address the causes. I have seen older patients who took sixteen different types of drugs, where one type of drug was merely used to suppress the side effects of one of the other drugs. Overmedicalisation is a serious problem in many western countries.

Of course, there are multiple reasons for this. One of the reasons is that it's easier and quicker for a doctor to pre-scribe a hypertension drug or a statin than to take half an hour to explain the importance of a healthy diet. Moreover, lots of people are not prepared to change their lifestyle in a certain way, and they just want a pill 'to make it go away'.

Personally, I think this is the heart of the problem: our

universities do not stress the importance of preventive medicine, and don't lecture their medical students about it, and neither do they stress the importance of wholesome food to our health, and in order to prevent chronic diseases. Because of this, as doctors, we cannot sufficiently inform and convince our patients. At best, we tell a patient that he 'needs to eat a lot of vegetables and fruit, and not too many fatty things'. This will not really convince a patient. But if you can tell your patient that some types of cancer occur 5 to 10 times less in areas where people eat more healthily, or that the chance of a heart attack is 45 percent less in people who eat a portion of walnuts every day, you will be able to convince and motivate a whole lot more patients. And then it's up to the patients to decide whether or not they want to make the effort to live healthier lives.

SUMMARY

Most medication does not heal but **suppresses the symptoms** of disease.

Many people underestimate the **side effects** of medication.

Even medication that is known to cause **few side effects** will still disrupt the equilibrium in the body. A few examples of these kinds of medication are gastric acid inhibitors (proton pump inhibitors), sleeping pills and painkillers.

Patients need to be **better informed** about the consequences of taking medication, and about the importance of a healthy diet to the body (with solid statistical examples, among other things).

Some of the reasons that this doesn't happen are lack of time, underestimating the intelligence or motivation of the patient, and a lack of training of health carers with respect to preventive medicine (there is too much emphasis on curing disease, instead of sustaining health).

5

Some insights regarding our health

Exercise? Or movement?
Doing exercise with the aim of losing weight is not a good idea. It just does not work. It is not exercise, but nutrition that is pivotal to losing weight. Suppose one eats two chocolate bars in one day, there will be an extra 500 kilocalories to burn and in order to convert those 500 kilocalories into energy, one must cycle for two hours. By simply eating just two more bonbons, one must walk one hour and 20 minutes in order to convert the extra calories into movement. Therefore, by simply not eating these in the first place you can save yourself half an afternoon of exercise.

Research also shows that exercising barely has any influence on weight and that what you eat is much more important, which applies even more for overweight people. A large study at Harvard University shows that physical exercise and sport are simply insufficient to help overweight people prevent further weight increase.[249] First of all, they must pay attention to what they eat. So, if you want to lose weight you are better off trading your gym membership for a good diet book.

Nevertheless, exercise is healthy; not so much to lose weight as to prevent a number of chronic disorders. Exercise reduces the chance of a heart attack, dementia, depression and stroke. Moreover, it is never too late to start doing exercise. A study that appeared in *The Lancet* showed that people who just start to exercise when they are middle-aged (and this is only two times per week) had a 62 percent reduction of Alzheimer's disease.[250] 62 percent! That is a number that would make a neurologist happy. Take a different disorder that occurs often in older people, especially when they have high blood pressure: mini-cerebral infarctions. These are small infarctions in the brain, caused by a small blood vessel that bursts. In contrast, a large cerebral infarction (a stroke) often causes sudden, conspicuous loss of function, which manifests in a speech disorder or a paralysed arm. But mini-cerebral infarctions are less conspicuous – they appear continuously and everywhere in the brain and do not cause any striking loss of function, yet they still produce more overall brain damage, which characterises itself through loss of concentration, forgetfulness, difficulty in recalling memories, slower thinking, etc. One study in *Neurology* showed that people who move on a regular basis have a 40 percent less chance of these mini-infarctions.[251]

Numerous studies also show that exercise helps against depression and that exercise is good for the mind. Exercise is at least as good in treating depression as antidepressants and is, according to some studies, even better.[252] I remember the story of a patient who had been depressive for many years. She had tried almost all types of antidepressant, including entire exotic cocktails, but nothing helped. One day she decided to go for a half-hour walk every evening, come rain or shine and whether she wanted to or not. She felt better after two weeks and after three months was relieved of the depression from which she had suffered for many years. People who exercise feel more motivated in the following days and have more desire to take things on. That is also logical because research shows that physical exercise causes the brain cells to secrete more neurotransmitters such as serotonin and dopamine. Exercise also makes the brain secrete various other substances that maintain and stimulate the brain cells. These substances help with the building up of extra nerve connections and enable brain cells to work better. That effect can even be seen on brain scanners. Walking just three times a week for 40 minutes increases the hippocampus, an area of the brain that is involved in the memory.[253] It is striking that the control group in the study, which did not walk and 'only' did weightlifting and stretching, showed no growth in the hippocampus. We will come back to this issue in a moment.

It is a big misconception to think that you must do hard exercise and tire yourself out in order to exercise 'sufficiently', or even worse, that you must sweat heavily and that your heart rate must be significantly stimulated. That is unnecessary. With physical exercise, it is not so much the intensity that is important, but the frequency. That means that you must do some form of regular exercise, which does not mean two hours of intensive running, swimming or basketball once per week, but preferably one half hour of walking four times per week. Walking is one of the healthiest sports a person can do. We have just seen that walking causes certain areas in the brain to become larger and that weightlifting does not have the same effect. That walking is healthy for the body has a long history. Mother Nature made people so they could walk. In prehistoric times, we caught our prey by walking after it. Imagine for a moment a prehistoric landscape, 30,000 years ago. Our ancestors stand before a flock of gazelles and choose one; they begin to walk after it. The gazelle looks up shocked and walks a couple of hundred metres away and continues to graze. But our ancestors do not give up and come walking in the distance. The gazelle looks up again, walks a couple of hundred metres further and then stops to graze again, although somewhat more nervous than previously. In the meantime our ancestors come persistently walking again in the distance. And so on. Until the gazelle falls down, exhausted from fatigue and stress and our ancestors have a piece of meat to eat; not by running after their prey but walking as a human. People are rather unique in the animal kingdom in terms of covering large distances at a continuous slow tempo. That 'slow tempo' may also be somewhat higher, such as speed walking. However, intensive exercise can be unhealthy and can even decrease the life-span,[254] which is not surprising. Top athletes do not always reach 100 years of age and often have a considerable beer belly when they reach their fifties because their mitochondria have been worn out through too much exercise.

I sometimes come across patients who look incredibly good for their age: 70 or 80 year-olds who look alert, full of life and twenty years younger and I cannot stop myself from asking them all sorts of questions about their lifestyle. Often they tell me that they walk frequently. I once came across an alert 89-year-old man who, apart from a herpes infection on his face, looked extraordinarily good. What that man had done since he retired was to gently exercise for half-an-hour each day on his exercise bike. In short, it is not so much exercise that is important, as it is to move and, above all, to keep moving.

A beautiful article in *Nature* shows the importance of physical exercise in preventing chronic diseases such as cardiovascular diseases and diabetes. Here you see a diagram from this study:

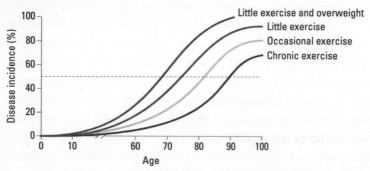

Source: Handschin, C. and Spiegelman, B.M. (2008), Nature 454:463–469

Chronic exercise means regular exercise. A 70-year-old who exercises regularly has an approximately 10 percent chance of a chronic illness, but someone of the same age who does not exercise has an approximately 40 percent chance of a chronic illness. This effect is even more pronounced when the non-exercising person is obese or has a beer belly. That person has five times as much risk of a chronic disease as the person who exercises regularly.[255]

We can ask ourselves why physical exercise is so good at preventing chronic disorders. One important reason for it is that physical exercise decreases inflammation. People inflame continuously and everywhere. With this I mean micro-inflammation – inflammation at the cellular and molecular level. There is continuous inflammation in our blood vessels, which cause the clogging of the blood vessels (arteriosclerosis). There is inflammation in our fat tissue, which can cause us to develop diabetes; there is inflammation in the brain and in other organs, which increases the chance of Alzheimer's and cancer. Through exercise, the micro-inflammation occurring all over the body becomes more suppressed, which protects against atherosclerosis, dementia, diabetes and cancer.

Indeed, exercise can even reduce the risk of cancer. Large studies show that women who exercise have a 49 to 79 percent lower chance of developing breast cancer, depending on the age and the type of breast cancer.[256] Men with prostate cancer, who regularly exercise, have 70 percent less chance of relapsing.[257] Dr Demark-Wahnfried, a researcher into breast cancer, set the level of risk reduction through physical exercise at the same level as Herceptin, a new and revolutionary drug against breast cancer that is considered 'as a turning point in the battle against suffering and death due to cancer' according to the words of Dr Von Eschenbach, Director of the American National Cancer Institute.[258]

But as already noted: regularity is the key. In order to have

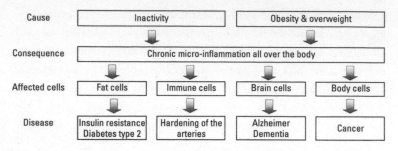

Cause	Inactivity		Obesity & overweight	
Consequence	Chronic micro-inflammation all over the body			
Affected cells	Fat cells	Immune cells	Brain cells	Body cells
Disease	Insulin resistance Diabetes type 2	Hardening of the arteries	Alzheimer Dementia	Cancer

Source: Handschin, C. and Spiegelman, B.M. (2008), Nature
454:463–469

an effect on breast cancer, patients must walk for half an hour at least six times per week, but for colon cancer this is doubled. However, if a cancer patient can at least halve the chance of a relapse through simple exercise then that is undoubtedly worth the effort.

In short, much trouble can be prevented by, for example, exercising just three times a week for twenty minutes. Making time to exercise is not easy, especially three times per week, but it can be planned, even in the busiest schedule. Try to exercise with friends so that you motivate each other. Set up a clear weekly schedule so that you know when you must exercise and so that your other activities can be planned around it. For those who have a smart phone: install a personal coach app. This is a small programme that, via a GPS in your mobile phone, can trace how many kilometres you have run, cycled or walked. Every few minutes, a voice can be programmed to tell you what your average speed is and how much distance you have already covered, plus a couple of compliments on your performance. These can be very motivating.

Some people buy an exercise bike or treadmill for their home; however, the experience of many exercise coaches and physicians is that most people abandon their exercise bike after a couple of weeks. After all, it is rather boring to pedal on a bicycle a few times each week while staring at a wall. This problem can be solved by placing your exercise bike in front of a TV or laptop so that you can watch a film with headphones on so that the dialogue does not become drowned out by the treading of your pedals. For others, an MP3 player, CD player or a radio is sufficient to prevent boredom setting in. You can do it anyway you like, as long as exercising at home is not boring.

SUMMARY

Exercising in order to lose weight does not work if nutrition is not addressed first.

Regular movement, in the form of **aerobic movement** (such as cycling, walking or swimming):
- reduces the chance of dementia, heart attacks, strokes and mini-strokes;
- works at least as well as antidepressants for most depressions;
- reduces micro-inflammation everywhere in the body;
- can reduce the chance or relapse of some cancers (such as breast cancer).

Walking at a brisk pace is one of the healthiest forms of moving.

It is not the intensity, but the **regularity**, of the exercise that is important. Moving three times per week for twenty minutes is the minimum.

In order to keep persevering:
- exercise with friends or family;
- plan your exercise activities beforehand and at fixed moments of time in your schedule;
- record your progress in a notebook or smart phone (number of hours exercised, number of kilometres covered, etc.).

Continued overleaf

Anaerobic exercises (such as weightlifting):
- stabilises mainly the sugar levels and makes the tissues more sensitive to insulin (as a result of which, the sugars will be removed faster from the bloodstream);
- inhibits sarcopenia (the decrease in muscle mass when becoming older).

Truly delaying ageing

The anti-ageing business is booming like never before. Growth hormone treatments and antioxidants must ensure that harmful free radicals can no longer do their awful work, which in turn, delays the ageing process. I have already explained in this book that on the contrary, growth hormones actually accelerate the ageing process and that (soundly performed) scientific research shows that antioxidants do not delay ageing.

Viewed scientifically, there is just one method that can significantly delay the ageing process; it involves our nutrition and this intervention is called calorie restriction. Calorie restriction (CR) consists of eating approximately 25 percent less food than would normally be required. If an adult female, for example, needs 2000 kilocalories, then she should eat just 1500 kilocalories daily when on calorie restriction. It is strange that eating less ensures that you live longer. You would say that light starvation makes you weaken faster and break down faster because the body does not get sufficient calories.

Calorie restriction was discovered by Clive McCay, a biochemist and researcher at Cornell University. He wanted to know what would happen if rats were given very little food for a large part of their lives. He expected that his rats would die sooner because they never got enough to eat, but on the contrary, the opposite was true. The lifespan of the rats increased, and the increase in the lifespan was proportional to the degree of calorie restriction. With solid calorie restriction, where the rats received less than half of what they normally need to eat, the lifespan increased from 1,000 days to 1,800 days, which corresponds to a human lifespan of 150 years![259] The rats also remained much healthier up to their old age: the risk of cancer was many times lower, as was the chance of diabetes, cardiovascular diseases and cognitive deterioration. In old age, these rats were still fit and their fur was shiny.

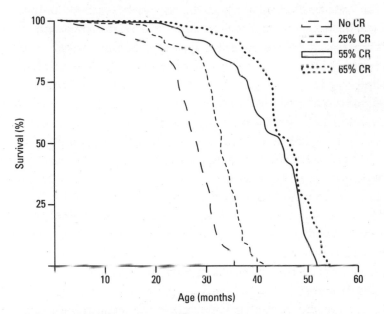

By eating less, organisms age less quickly. At the time that most mice that were following a normal diet would be dead, the majority of the calorie-restricted mice were still living.
Source: The retardation of ageing in mice by dietary restriction: longevity, cancer, immunity and lifetime energy intake.
Journal of Nutrition, *1986*

Calorie restriction also has an important effect on the ageing process in animal species more closely related to us than rats. The first results of a large study investigating the effect of calorie restriction in rhesus monkeys recently appeared in Science. For 20 years, these monkeys ate 30 percent less than 'normal'. The results were spectacular. After 20 years, 37 percent of the non-calorie restriction monkeys had already died, while only 13 percent of the calorie restriction monkeys had died. In short, the mortality in the rhesus monkeys on calorie restriction was three times lower. The monkeys under calorie restriction also had much less trouble with symptoms of ageing such as loss of muscle mass, cancer, cardiovascular diseases and the shrinking of brain volume. The most spectacular was the influence of calorie restriction on diabetes. Of the monkeys that could eat as much as they wanted, 40 percent developed diabetes or pre-diabetes, while none of the CR monkeys developed problems with their sugar metabolism.[260]

Not long after this study was published in *Science*, another study in *Nature* stated that calorie restriction didn't affect the lifespan of Rhesus monkeys.[261] However, it is important to view the bigger picture: the number of studies that show that calorie restriction lengthens life expectancy is much larger than that of studies that don't show any effect, or indicate a negative effect. Moreover, there was an important difference between the study that was published in *Nature* (that didn't show an effect of calorie restriction) and the study in *Science*. The monkeys in the comparison group (this is the group to which the monkeys that were subjected to calorie restriction was compared) were allowed to eat as much as they wanted. Just like humans, you could say. However, the comparison group in the *Nature* study was fed a fixed amount of calories per day. They could never eat as much as they wanted. This might explain why there was no noticeable difference between the lifespan of the comparison group and that of the CR-group in the *Nature* study, since the comparison group also experienced a form of calorie restriction.

This also explains why the comparison group of the *Nature* study barely contained older monkeys which were overweight or suffered from diabetes, while the *Science* study comparison group exhibited old monkeys that in the majority of cases suffered from being overweight or (pre-) diabetes and early death; these were the monkeys that had been allowed to eat as much as they pleased, throughout their entire life. A bit like humans, who can also eat as much as they want, and who often show signs of overweight and diabetes when they reach old age.

It's also important to note that even the *Nature* study demonstrated that calorie restriction has a beneficial effect on the health of older monkeys: their blood sugar levels were better and they had a smaller chance of getting cancer and less rapid ageing of the brain.

Most studies show that calorie restriction increases the

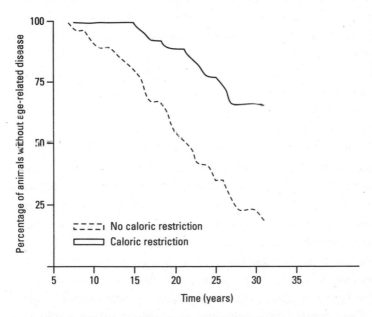

A moderate calorie restriction (consuming 30 percent fewer calories) drastically reduces the chance of ageing diseases in rhesus monkeys. *Source: Caloric restriction delays disease onset and mortality in rhesus monkeys*, Science, 2009

lifespan of various animal species. It is very interesting to see that calorie restriction increases the *maximum* lifespan as well. There is, after all, a big difference between a substance that increases the *average* lifespan or the *maximum* lifespan. The average lifespan for a human is approximately 80 years. The maximum lifespan for a human is approximately 120 years. The maximum lifespan is mainly determined by genes (and calorie restriction works on the genetic level by regulating transcription factors and the like), while the average lifespan is mainly determined through our life habits. Someone who, for example, takes vitamin C supplements will perhaps be able to extend his average lifespan somewhat (certainly if he has a vitamin C deficiency), but vitamin C will not extend the maximum lifespan. That is an important nuance, because after all, vitamin C does not delay true ageing.

Substances or interventions that extend the maximum lifespan are substances that delay true ageing. There are very few substances or interventions known in science that increase the maximum lifespan, such as calorie restriction or rapamycin (the substance that made mice live longer in a previous chapter by inhibiting protein production).

How can calorie restriction now delay ageing? If you eat less than 'normal' via calorie restriction, the body is going to think that it is in a difficult environment, with scarce food sources and a danger of dying of starvation. Therefore, the body is going to slow down its growth processes because growth costs food and there does not appear to be enough food present. The body will not only stop growth, but will also invest more effort in the maintenance and repair of cells. This occurs because food is scarce and the body must continue to operate with the same cells, so they are better maintained.

Calorie restriction therefore inhibits growth and promotes maintenance. We have already seen in this book many times how growth makes us age faster. Growth in adults is the constant production and replacement of cells, proteins, DNA, etc. If the growth is inhibited through calorie restriction, then we age less quickly. In contrast, if we eat as we are accustomed to in our western society, with an overabundance of food, then the body will no longer go into safety mode; in fact, it takes the brakes off: plenty of proteins are created, the mitochondria (the energy power plants of our cells) run at full speed, creating many free radicals, hormones are created and secreted in abundance, etc. All of this rash cellular activity causes the body to age faster.

In the US, there are various groups established by people who practise calorie restriction. I have met many people who follow calorie restriction and these people look strikingly younger than their age. Moreover, they have much more healthy biomarkers: lower blood pressure, less inflammation in their body in general, a higher, healthy HDL cholesterol level, a better handling of glucose and insulin etc. All their parameters are comparable with those of younger people. Medical history also has numerous people who became extremely old and it was often that these people never ate too much.

The earlier someone begins with calorie restriction, the longer that person can live. But it is never too late to start calorie restriction. Dr Roy Walford, a renowned researcher into calorie restriction, estimates that when someone begins with calorie restriction at the age of 50 to 60 years, he can still live 10 to 15 years longer and be plagued for fewer years by chronic diseases. That has been shown in a study conducted by Dr Eduardo Vallejo, where 180 people aged 65+ were divided into two groups. One group received a standard diet of 2,300 kilocalories per day, while the other group received a standard diet of 2,300 kilocalories on one day and a diet of 885 kilocalories on the other. As a result, the latter group ate an average of 1,590 kilocalories per day, which corresponds to an average calorie restriction of 30 percent. The study lasted three years and the elderly people who participated in the calorie restriction diet were, on average, in the hospital 50 percent less and their mortality was halved when compared to the people who had consumed a normal diet with 'sufficient calories'.[262]

This form of calorie restriction is called ADF ('alternate day fasting'), but is difficult to sustain in the long term. A more bearable method is partly fasting for two days a week. These can be consecutive days, or separate days. During these 'fast' days, test subjects eat one or two small meals a day, no more than 600 kilocalories in total (for example, a breakfast and a dinner that each contain 300 kilocalories). Professor Mark Mattson, an expert in the field of (brain) ageing, suspects that it would be even better to just eat a single meal of 600 kilocalories a day, instead of two individual meals of 300 kilocalories each.[263]

In short, partly fasting is a kind of temporary calorie restriction that can be beneficial to your health, according to recent scientific studies.

Of course, it is important that people do not become deficient in essential nutrients with calorie restriction so they must always consume sufficient nutrients such as vitamins and minerals. This concept is called CRON: *calorie restriction under optimal nutrition.* Furthermore, calorie restriction is not suitable for everyone. Calorie restriction is not advised for pregnant women, children (who must still grow) and people with serious diseases such as amyotrophic lateral sclerosis (ALS, a neurological disease). That is why it is recommended you always ask your doctor for advice and go in for a checkup if you intend to change your eating habits in a major way.

For many people, calorie restriction is a bridge too far. They think that people who practise calorie restriction starve themselves and walk around constantly with a rumbling stomach. However, that is not the case: you just eat less, approximately 25 percent. The body quickly adjusts itself to this new habit and soon people no longer have feelings of hunger when they consume a smaller meal.

The food hourglass proposed in this book rests on the concept of calorie restriction. The basis of the hourglass is, after all, vegetables. Vegetables contain a great many nutrients and very few calories. Nutrients are healthy substances such as vitamins, minerals and flavonoids. Therefore, whoever eats many vegetables will consume a higher number of healthy substances as well as fewer calories. The opposite is true when someone eats junk food, which contains a great many calories but very few nutrients. A healthy meal automatically contains fewer calories so you are already practising a kind of calorie restriction.

In addition, vegetables are very high in fibre. As a result, they fill the stomach faster and also leave the stomach slower than unhealthy food. As a consequence, those who eat healthily feel satisfied faster even though they consume fewer calories. A study at the University of Alabama shows that people who follow a healthy diet that consists mainly of vegetables, fibre and fruit are already full after consuming just 1500 kilocalories, while when these test subjects ate meals that consisted of industrially prepared food with many sugars and fats, they were only full after eating 3,000 kilocalories![264] In short, whoever eats healthy food, automatically performs a kind of calorie restriction.

The composition of healthy food ensures that in exchange for few calories, many healthy nutrients and fibre are consumed. As a result, you lose weight, even if you may already eat as much as you want from the food hourglass. People who are constantly busy counting calories are actually on the wrong track.

Moreover, you can reduce your calorie intake by 25 percent quite quickly. For example, by eating one less meal per day, one-third fewer calories will be consumed. This means that, for example, the evening meal can be dispensed with or at least replaced by a very light meal. Many people, however, eat the largest hot meal of the day in the evening. That custom is discouraged by many nutrition specialists. All of that food is in the stomach when you go to sleep so you are more likely to be bothered by acid reflux. Moreover, all of that food must be digested while you sleep. We do not move during our sleep and all the calories that have been ingested are not converted into movement but stored as fat. Digesting a meal also costs the body a lot of energy. Normally, one litre of blood passes through the gut every minute but after a meal this can increase to four litres per minute! (The heart pumps around approximately five to six litres of blood per minute.) This enormous supply of blood is required to provide enough energy for the intestines and to absorb nutrients from the gut.

At the time that you are sleeping this energy could be better spent on the repair and maintenance processes of the body, rather than on the digestion of a rich evening meal. Ideally, the largest meal would be consumed in the morning, with a smaller meal in the afternoon and a light meal in the evening or even no meal (at most, some easily digested fruit or a quick bowl of thick soup in order to hold back the feeling of hunger). The saying, 'Eat like a king in the morning, a prince in the afternoon and a pauper in the evening,' hits the nail squarely on the head. No longer having to cook extensively in the evening not only saves time, it is also good for the health.

And it can bear fruit. For example, a man who had lost a lot of weight came to see me for an appointment. I read in his file that he had lost 40 kilos and I asked him what he had done in order to lose this weight. 'Nothing,' he answered, 'I just stopped eating an evening meal, just some fruit and such.' I read the opposite in an interview with the corpulent ex-politician and ex-stock exchange guru, Jean-Pierre van Rossem, who cannot hide his frustration about not being able to lose any weight: 'I only eat one full meal in the evening. During the day, I eat little containers of lemon yoghurt, about ten per day, and I drink litres of Coca-Cola Zero – but I still weigh 110 kg!' Apart from the fact that Coca-Cola Zero is full of artificial sweeteners and phosphates, and yoghurt is not good for the gut, the heavy evening meal ensures that the kilos do anything but melt away at night.

One more thing, I sometimes hear people say that being slightly overweight is healthy. After all, studies from insurance companies and scientific research show that persons who are slightly overweight have a lower mortality rate. People often conclude from this statement that being slightly overweight is healthy but that is the wrong conclusion. After all, the group that contains thin people also contains more sick people. People who have cancer, AIDS or tuberculosis are usually thin because they lose a lot of weight due to these diseases. Smokers also weigh less on average because they lose weight due to smoking. Therefore, the group of thin people also contains unhealthy people who increase the average mortality of the group. As a result, it seems like thin people do not live as long, which is not the case.

SUMMARY

Calorie restriction is the only method which is scientifically proven to be able to significantly delay the ageing process.

With calorie restriction, you eat approximately 25 percent fewer calories than 'normally' required.

As a result, the body goes into an **energy saving state** and:
- fewer proteins, hormones and other substances are created;
- the cells are better maintained.

Continued overleaf

Healthy nutrition is a kind of calorie restriction because healthy nutrition:
- contains few calories and many nutrients (vitamins, minerals, flavonoids);
- contains much fibre that fills the stomach faster so that you eat less;
- often contain substances that cause the metabolism to run faster (such as EGCG in green tea, omega-3 fatty acids in fish or iodine in seaweed).

By eating a light (or no) meal **in the evening**:
- the daily calorie intake is reduced considerably;
- the body needs to invest less energy at night in digestion and more energy remains for bodily repair processes;
- the energy gained from the food is not immediately stored in the form of fat;
- the chance of acid reflux is reduced.

It is a misconception that **being slightly overweight** is healthy.

Mind and body

When I was an intern in the cardiology department (cardio-vascular diseases), it struck me that most of the patients who had suffered a heart attack were either too fat or too thin. One group of patients had too much belly fat, while the others were simply too thin. The thin people usually gave a nervous or anxious impression and their nightstand was often full of sleep medication and sedatives. It is no secret in medical literature that too much stomach fat, as well as chronic stress, significantly increases the chance of a heart attack.

Belly fat (visceral or abdominal fat) is fat that accumulates in the belly: also well known as a 'belly' or 'beer belly'. Medically speaking, belly fat is much unhealthier than fat that accumulates in other places around in the body, such as in the buttocks and thighs. Physicians call that type of fat 'subcutaneous' fat. People with a lot of belly fat are often shaped like an apple whereas people with a lot of subcutaneous fat look more like a pear (due to the fat accumulation on the buttocks and the thighs).

In short, two people can weigh the same and have the same amount of fat in their body, but the apple-shaped person with a lot of belly fat runs a much higher risk of a heart attack than the person who has fat mainly in the buttocks and the thighs. The reason for this is that belly fat secretes numerous inflammatory agents that are extremely unhealthy. These inflammatory agents (*adipokines*) clog the blood vessels, cause the body to respond more poorly to insulin (so that the chance of diabetes increases drastically), promote micro-inflammation everywhere and can even damage the brain. People with a lot of belly fat have almost three times as much chance of dementia.[265] In this way, you can see how the belly can influence the brain. But the

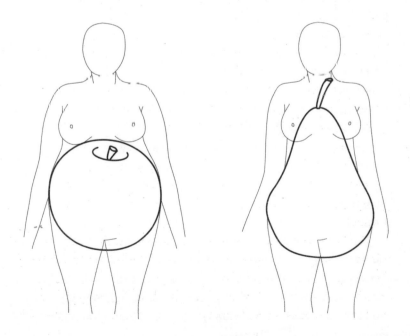

opposite also applies: the brain can also strongly influence the body.

The second type of patient in the cardiology department was the thin, nervous or anxious type. After all, stress, anxiety and anger have a great influence on our health. This influence is naturally underestimated because stress and anger are immaterial things and it is difficult to imagine how immaterial things can influence our tangible body. But the connection between mind and body is clear. According to one study that appeared in the medical journal, *Circulation*, it appears that people who easily become angry by nature and who lose their temper faster (type A personalities) have three times more risk of a heart attack.[266] Those who experience constant stress have 2.5 times more chance of a heart attack.[267] Psychiatrists have known for a long time that people with depression are much more susceptible to cardiovascular diseases. One study in women who had had a heart attack showed that women who experienced depression afterwards, or were socially isolated, had four times more chance of a new heart attack than women who were happier.[268] Scientists agree that depression is just as great a risk factor to cardiovascular diseases as smoking or diabetes![269]

What these studies show is that the psyche can strongly influence our health. Mind and body are not separated. It is wrong to make a distinction between mind and body like the French philosopher Descartes, because the mind is a product of the body (specifically, of our brain) and the body is strongly influenced by the brain. Someone who is regularly anxious or depressed is much more susceptible to heart attacks, just like someone who does little exercise or has an unhealthy diet (and therefore has an unhealthy body) has a greater chance of depression, anxiety disorders and even schizophrenia.

Our so-called 'immaterial' mind and the 'tangible' body interact with each other via numerous different mechanisms.

When we experience stress the body secretes the hormone *cortisol*. Cortisol readies the body for 'the fight' or 'the flight'. Previously, a dangerous situation was one in which a sabre-toothed tiger attacked you but now, for example, it is when we have to make a PowerPoint presentation under a tight deadline. Cortisol increases the blood pressure and makes the blood more 'coagulable'. After all, increased blood pressure ensures more blood flow to the limbs and muscles to help us fight or flee. By making the blood more coagulable, we will not bleed too heavily when a predator opens a wound in our arm or a leg. This was perhaps all very useful in prehistoric times but now cortisol ensures that higher blood pressure and more coagulable blood significantly increase our chance of a heart attack.

Cortisol also suppresses the immune system and as a result, we become more susceptible to infections and even cancer because the immune system, among other things, cleans up cancer cells. In a famous study that appeared in the journal, *Science*, mice were injected with tumour cells in such a dosage that they cause cancer in 50 percent of the cases. The mice were divided into three groups: one group of mice received an electric shock every so often (a stressful occurrence), a different group of mice also received an electric shock every so often, but could immediately turn it off by pushing on a lever and a control group did not receive any electric shocks. The group of mice that could stop the electric shocks suffered stress, but they had a feeling of control over it. Stress is, after all, worst when you have no feeling of control over the situation and you no longer have your fate in your own hands.

The results were astounding. Over the course of time, only 23 percent of the mice that received electric shocks were still alive; of the mice that received electric shocks, but could control them, 64 percent were still alive; of the control group, half of the mice died of the cancer, as predicted.[270] The mice that were exposed to uncontrollable stress had almost three times more chance of dying of the cancer compared to the mice that had a feeling of control.

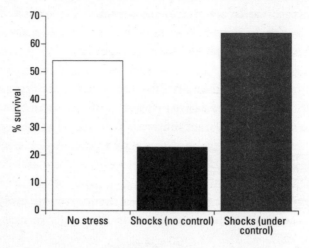

Mice die faster from stress. *Source: Tumor rejection in rats after inescapable or escapable shock*, Science, 1982

We now know that stress has an important influence on our health. But it is not easy to live in a more calm and relaxed manner in this hectic world. Nevertheless, there are numerous relaxation methods that help us to do so, such as meditation, yoga, self-hypnosis or breathing techniques. People who practise meditation have a stronger immune system (and are less sick) and have lower blood pressure. Meditation can significantly reduce the chance of heart attacks and strokes. According to a study in *Circulation,* in which 201 people participated, it appeared that the total mortality of those who meditated for 20 minutes twice daily was at least 43 percent lower.[271] The well-known psychiatrist, Dr Norman Rosenthal, said in the context of this study that 'if meditation was a medicine, it would be a *billion-dollar blockbuster*'.

Relaxation exercises not only relax you in the hours

following the exercise. After only a few weeks, you will also feel yourself more relaxed throughout the day, even without first doing relaxation exercises. It has been shown that meditation can bring about permanent changes in the brain. In this way, regular meditation exercises cause specific areas in the cerebral cortex to become thicker, so thick that scientists can even detect this brain scanners.[272]

There are many different forms of meditation, but all of these forms of meditation can be roughly divided into two groups: closed and open meditation.

In closed meditation, people concentrate on a point, which can be a spot on a wall, or a repetitive noise that they may or may not make themselves, or a specific object in their imagination. In open meditation, such as *mindfulness*, people

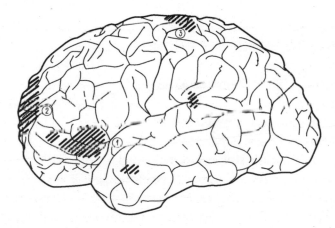

Some areas of the cerebral cortex that become thicker when meditating:
1. Insula
2. Brodmann 9 and 10
3. Somatosensory cortex
Source: Meditation experience is associated with increased cortical thickness, Neuroreport, 2005 *(with permission from Dr Sara Lazar)*

observe their thoughts unbiased and let go of them. Although mindfulness has become very popular in recent years, it is my impression that closed meditation is more powerful. Certainly, if you lie in bed or sit in a chair and stare for a few minutes at a point that is just above your field of vision (so that you must stare a little bit upwards), this activates many other neural networks than someone who, for example, looks downwards or just stares straight ahead of themselves.[273] When people stare upwards for a few minutes, and then close their eyes and still keep staring, they can then get the impression that they are looking at a point that is in the middle of their forehead. This point is called the 'third eye' in the Far East and it is over this, on the skin of the forehead, that the famous Hindu dot is painted. This technique activates specific neural networks that generate our body image and people can end up in a trance after some time.

You can also do yoga in addition to meditation. Self-hypnosis is another way in which you can relax and this is more individual. It consists of sitting relaxed in a chair or lying on a bed at home, whether or not staring at the 'third eye', and listening to a CD with a soft voice that whispers auto suggestions to you such as: 'breathe in and out deeply and feel how your chest goes up and down slowly,' or 'feel how your eyelids slowly but surely become heavy,' or 'how a warm feeling arises in your fingertips,' etc.

Breathing also plays an important role in relaxation, as well as anxiety. Many people breathe incorrectly. It is important to breathe through the abdomen and not through the chest. Too many people breathe through the chest, which is unnatural; normally, people only breathe through the chest when they make a heavy effort. But when they continuously breathe via the chest and in the meantime, just sit on a chair, receptors in the chest send incorrect signals to the brain. The brain interprets these signals as, 'Attention, chest breathing is occurring so a heavy effort is being made here.' The brain thinks that the body is in a state of stress and danger, through which people will also feel more stressed and anxious. Whoever continuously breathes through the chest, is going to spontaneously feel more nervous and more stressed. People also have a greater tendency to hyperventilate through this chest breathing. Sometimes the muscles between the ribs can even become irritated, so that a person gets chest pain. I have often seen patients in hospital who thought they had had a heart attack when in fact they were just breathing through their chest and hyperventilating with chest pain, lying there, surrounded by an ECG device, with nurses and a doctor telling them that actually nothing is wrong.

Thus, good breathing occurs via the abdomen and occurs slowly: at least five counts breathing in and five counts breathing out. One of the best ways to get stress and feelings of anxiety under control is by sitting up straight in a chair, with one hand on the stomach and one hand on the chest (that may not move up and down), and then to breathe in and out through the abdomen ten times slowly. The body and the brain will interpret this new breathing pattern and automatically ensure that we feel calmer. Currently, there are many studies appearing about the neural pathways that run between the lungs, the heart and the brain. These studies show that people who breathe more calmly also have a better heart rate and as a result, have a lower chance of dying.[274]

Together with good breathing, good body posture is also of great importance. Perhaps this sounds strange, but as said earlier, we no longer live naturally. We eat unhealthy food; we barely move; we live inside, deprived of sunlight; and we spend the greatest part of our lives sitting down. As a result of this, our body posture automatically changes; specific muscle groups weaken; and tendons and ligaments become stiffer, through which we become more susceptible to back and neck problems. And that's not all.

Poor body posture can result in a number of negative effects on the body that should not be underestimated. A curved lower back, hunched shoulders or a poorly positioned neck can press on many kinds of nerves, blood vessels and organs in the body, which can cause various chronic complaints, such as fatigue, headache, stomach pain, shortness of breath, neck pain, back pain and even acid reflux. I remember a young person in his twenties who had already suffered for more than five years with heartburn, muscle pain and fatigue. He exercised fervently, but had to stop due to fatigue. He had lost ten kilos in that time because he could barely eat anything – he got painful acid reflux after every meal. He visited all sorts of doctors and professors, from gastroenterologists to endocrinologists and general practitioners. Antacids, gastric emptying accelerators and other medications offered little relief. The young man was tall and thin, with a typical gangly, teenager-like posture. I advised him to go to a physical therapist for exercises that strengthen the back muscles and hip muscles, so that his back would be pulled to the back and his pressed-in rib cage would be pulled open. After a few exercise sessions he already felt noticeably better, and he was practically relieved of his symptoms after a few weeks. His hunched-over body posture probably caused, among other things, the diaphragm to be restricted, through which the blood vessels and nerves that run toward the stomach became clamped off, which made the stomach sick. The young man's back was also too straight, which may have contributed to these symptoms (the back normally has two natural curves and a gentle s-shape). I have still not found a similar case in medical literature, so in the event that this 'gastrointestinal reflux and excessive chronic fatigue caused by poor postural body posture or straight back syndrome' should occur again, then I would like it to be called Verburgh syndrome. Of course, this is a serious (and hopefully rare) case, but it illustrates what kind of consequences a poor body posture can have on the body.

We often underestimate our banal daily habits, such as food, breathing or body posture. But precisely because these issues are so daily and constant, they can continually undermine our health. With all the corresponding consequences.

SUMMARY

Not all overweight people have an equally large chance of a heart attack. There are two forms of overweight:
- 'apple-shapes' have a lot of belly fat;
- 'pear-shapes' have a lot of subcutaneous fat (including the thigh and bottom areas).

Belly fat is unhealthy because it:
- secretes inflammatory agents;
- drastically increases the chance of cardiovascular diseases, diabetes, depression and dementia.

The chance of a heart attack is increased by:
- excess belly fat (usually in overweight people);
- excess stress (usually in thin and stressed or anxious people).

Stress causes:
- an increased generation of inflammatory agents that cause the blood vessels to clog faster;
- an increased production of cortisol, which makes the blood more coagulable and suppresses the immune system.

The total mortality in people who meditate regularly can be significantly reduced.

Good breathing (through the abdomen and the nose) and good body posture are important for the general health.

It does not work without others

People are social beings. Locking someone up in an isolation cell is one of the cruellest punishments that you can imagine. I know of people who became psychotic from loneliness. People who are lonely will even secrete specific substances in the brain or activate genes that scientists can trace.

For millions of years, our ancestors have always lived in tribes, through which we have developed comprehensive social abilities, such as empathy and language. The part of our eyes around our iris even became white (the 'white of the eye') so that we can understand even better the non-verbal communication signals of our kind (monkeys have no white of the eye: theirs is black).

As social beings, it is of great importance to our health to remain in contact with our fellow humans – to be social. Numerous studies show that people who are social remain healthy longer and have less chance of dying. 'Social' does not mean, of course, that you have to hang around in a café every day or go with friends from one party to another, but that you should spend time with, and pay attention to your partner, friends, family and society in general. The latter can mean volunteer work, or the establishment of an association or serving a good purpose. This all comes down to the same thing: the serving of 'a higher purpose' that goes beyond your own self.

Even being a tiny bit social can have a significant effect: the mortality of elderly people who had a plant that they had to take care of was 50 percent lower than those who had a plant that they did not have to take care of themselves.[275] Giving a plant water can therefore make you live longer. An article in the *American Journal of Cardiology* showed that people who had a heart attack and had a pet had six times less chance of dying compared to people who did not have a pet.[276] Sometimes it is good for a physician to prescribe a dog or a plant.

The possibility to talk with others and to share emotions even has an influence on the progression of cancer. A study in *The Lancet* showed that women with breast cancer who were in a self-help group survived twice as long as women who were not in a discussion group.[277] Much turmoil was generated regarding this study. Some new studies, after all, showed no effect of discussion groups on cancer. This could be explained because the women in these studies were also not psychologically happier from these discussion groups, so that it is also logical that no physical changes appeared. More recent studies showed, however, that the effect of discussion groups applies mainly for women who suffer from a specific form of breast cancer, in particular, the oestrogen-receptor-negative breast cancer. With this type of breast cancer, researchers even saw a tripling of the survival time of women who were members of a discussion group.[278]

What this and a number of other studies show is that people are social beings and need others in such a way that their physical health depends on it. However, in these times of mobile telephones, email, hundreds of friends on Facebook and improved transportation possibilities, it is ironic that people feel lonelier than ever.

SUMMARY

People are **social beings** who need each other.

Caring for a plant or pet or participating in a discussion group can significantly reduce the risk of cancer or a heart attack.

And finally: a few extra tips on losing weight

The food hourglass is not a diet, but a healthy way of life, a way of life with the goal of delaying the ageing process; weight loss is a secondary benefit. But as with most diets, the application of the food hourglass incorporates a change in eating patterns. For some, this will even be a total change and given that people are creatures of habit, it is not easy to apply such a change. Moreover, all those addictive sugars and fats do not make it any easier, so here are a few tips that can make a diet successful.

1. Remove temptation
Throw all the crisps, biscuits (and any other sugar-rich food that makes your mouth water when you think about) into the bin. Make sure your fridge and cupboards are full of healthy snacks such as grapes, berries, nuts, apples, strawberries, dark chocolate, oatmeal, etc. If there are no bags of crisps, sweets or cake in the house you have no other choice than to resort to a bowl of strawberries, oatmeal or walnuts when you get hungry.

2. Eat many small snacks
Hunger causes people to make irrational choices and to reach for unhealthy sugar-rich foods, even though they know better. After all, hunger makes people's mouths water when they think of a delicious chocolate bar or a bag of crisps. However, a lot of munching can be avoided by preventing hunger and the best way to do that is to continually eat healthy snacks so that you never really get hungry. For example, eat an apple with a piece of dark chocolate every day at 10.30am and blueberries with a bowl of oatmeal at 3pm. In doing this, you can avoid suddenly getting hungry and avoid the craving for something sweet and unhealthy.

Moreover, eating a lot of healthy snacks ensures the metabolism keeps running. The stomach and gut continue digesting for the entire day, which costs energy so you can lose even more weight. Some go even further and abandon the three main meals and replace them with five or six smaller meals. Just keep the metabolism running is what I would say!

3. Eat a small meal in the evening and a substantial breakfast in the morning

Eat a light meal in the evening. This will help you to sleep better, the body will store fewer unnecessary calories and less energy will be invested in digestion and more will be spent on maintenance processes.

On the other hand, a substantial breakfast at the beginning of the day is required to set your body up for all the food that will follow during the day. People who do not eat breakfast in the morning get fatter more quickly than people who eat breakfast and consume the same number of daily calories.

4. Know why you eat

Some people do not eat just to satisfy their hunger, they eat out of habit; for example, ice cream in front of the evening TV, or a little container of yoghurt after each meal. It is important to place question marks against these habits. Breaking these habits is sometimes difficult but if they are replaced by new, good habits, you will be able to worry less for many years to come.

Other people eat to escape specific problems. They eat due to stress, because they are unhappy, out of loneliness or to escape certain anxieties. If this is the case, it is important to figure out for yourself why you are stressed, anxious or unhappy. If necessary, write these reasons down with possible solutions. If these problems are too serious, it important to seek help with a professional care provider.

5. Make an emergency kit

I call this an FAH kit: First Aid when Hungry kit. This kit consists of healthy snacks that can be consumed when you are really hungry and yearn for something sweet. You can keep this kit in your car, handbag or desk. An FAH kit within reach is necessary because in our society there is often no healthy food in the vicinity. Just imagine you get hungry in a train station where only hotdog stands and candy machines are located. The FAH kit ensures that you do not have to reach for fat and sugar-rich calorie bombs without any nutrients. An FAH kit can be simple: an apple in a handbag, a banana in a backpack or a little packet of nuts on your desk.

6. Keep a food diary

Keeping a food diary is a simple form of neurofeedback. For example, you can make note of your weight in a food diary together with the date so that you can follow your progress (or regression). If you can then see for yourself how you lose weight month after month, this will motivate you even more. You can also write down thoughts and feelings about food and your body in a food diary. Relapses can be recorded in the food diary, together with the reason why and how to help prevent them in the future. From this you can learn a lot and draw motivation. You can also note in a food diary what you eat on a daily basis so that you get an understanding of your eating patterns. It is even possible to keep a digital food diary, for example, in an Excel document or via an online health programme or weight loss app, where you can keep track of your weight on a daily basis with graphs, milestones to be reached, tips, etc.

7. Eat slowly and consciously

Proper digestion begins in the mouth. Everyone has six sturdy salivary glands around their mouth and these glands daily secrete 1½ litres of saliva that is full of digestive enzymes. A person should, on average, chew 30 times on each bite. In this world, full of temptations, possibilities and obligations, chewing slowly is not an obvious choice, but poorly digested food means nutrients cannot be absorbed so effectively and this places an extra burden on the gut.

Moreover, eating slowly ensures your body has enough time to create substances that provide a feeling of satiation. When someone eats slowly, that person allows appetite-suppressants to be secreted, which reduce the desire to eat further. On the other hand, if that person wolfs down his entire plate in less than 15 minutes, he will still feel hungry because the body has not had enough time to secrete the appetite-suppressants.

Eating slowly is indeed boring. In order to avoid becoming bored with all of that chewing, you can try to eat 'mindfully': enjoy the taste; concentrate on the texture of the food in your mouth; enjoy the colours of the food on your plate; think about the route that the tomato took before ending up on your plate, from the moment it was no more than a seed planted in the ground by a labourer in a large tomato plantation in Spain to the moment that you are now eating it with relish. Some philosophical souls can go one step further and when chewing, wonder about how the atoms of which their tomato consists were found once in the interior of giant stars that exploded billions of years ago. Indeed, the everyday is filled with scientific miracles!

Conclusion

New diets are introduced as regularly as clockwork. Some diets are quickly forgotten, others develop into hypes, which are then replaced by new proposals from the next batch of diet gurus.

Many scientists and nutrition experts are opposed to these diets. They are unhealthy because most of them concentrate on calorie groups. There are three calorie groups: carbohydrates (sugars), fats and proteins. Many diets label specific calorie groups as 'good' or 'bad'. According to the classic Atkins diet, you must primarily omit carbohydrates, while fats and proteins may be à volonté (unlimited). According to the paleo diet, you should get most of your proteins in the form of meat. According to more classic diets, you must primarily eat low-fat food. These diets, which each time exclude certain calorie groups and favour other calorie groups, are bad by definition. It is wrong to say that certain calorie groups are good or bad. It is not the calorie group itself that is important, but the form in which a specific calorie group is eaten. Thinking about it in this way, carbohydrates are healthy as well as unhealthy. Carbohydrates are unhealthy in the form of biscuits, white bread or potatoes, but they are healthy in the form of fruit and legumes.

Fats are healthy, as well as unhealthy. Fats are unhealthy in the form of trans fats and various saturated fats such as in the case of deep-fried foods, cake and butter, but they are especially healthy when they are packed in fatty fish and nuts. Proteins are healthy, as well as unhealthy. They are bad for our health in their animal form, such as in the case of red meat, but they are healthy when they are provided in the form of vegetable proteins in vegetables or Quorn.

In short, all the calorie groups are important. Try not to avoid 'the fats' or to neglect carbohydrates 'the Atkins way', but concentrate on the type of fats, sugars and proteins. All three calorie groups must be included. The food hourglass shows how these calorie groups can be provided in a healthy way.

A different problem is that many diets consist of different phases, such as the induction phase, attack phase, consolidation phase, maintenance phase, and so on. One reason why diet gurus subdivide their popular diets into such phases is to give their diet a somewhat more scientific élan. And because a healthy eating pattern cannot be patented. That is why they think up a lot of complicated phases in order to give their diet its own, personal tint. Moreover, they want to achieve a quick weight loss in the first phase. That occurs in the 'attack phase' and gets people to think, *Hey, that diet works really well because I lost eight pounds in two weeks.* But a too-rapid weight loss is unhealthy. Apart from the fact that this quick weight loss is often achieved by eating a lot of proteins, by definition it is bad for the body to lose weight quickly.

First, research shows that rapid weight loss shortens the lifespan of test animals such as rats and mice.[279] Of course, we will probably never be able to find out if this is also applicable for people because such research would take decades and cost enormous amounts of money. But according to many investigators, too-rapid weight loss causes the body to become damaged at the metabolic level, so that, in the longer term, the lifespan is shorter.[280]

Another reason why rapid weight loss is not healthy is that many toxic substances from the shrinking mass of fat end up in the bloodstream. After all, numerous toxic substances such as pesticides, dioxins and PCBs are fat-soluble substances that are stored in body fat where they can do little harm. But when someone loses weight quickly, these substances leak out of the shrinking mass of fat into the bloodstream where they can cause unhealthy side effects in the body. This process was observed in detail for the first time in test subjects who resided for years in *Biosphere 2*, a gigantic conservatory that was completely closed off from the outside world (including oxygen and food). The purpose of the *Biosphere 2* project was to see if a small group of people could survive for a long time in total isolation, which can provide interesting data in connection with the colonisation of other planets or making long trips in large spaceships. The eight scientists who lived in *Biosphere 2* ate mainly vegetables, nuts, fruit and seeds; of course they lost weight, and spectacularly because there was also too little food. But while their weight quickly dropped, the quantities

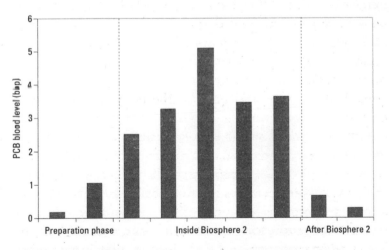

The level of toxic PCBs rose in test subjects in Biosphere 2 because they lost weight too quickly. *Source: Physiologic changes in humans subjected to severe, selective calorie restriction for two years in Biosphere 2: health, ageing and toxicological perspectives*, Toxicological Sciences, *1999*

of toxic substances in their blood rose significantly. The weight loss was too rapid and caused toxins to be quickly discharged from the fat into the bloodstream.[281]

Thus, there are various reasons not to lose weight too quickly and that is why physicians recommend that weight loss should be spread over a period of at least six to nine months, ideally over one or two years. Six months is the minimum, which is, after all, the minimum amount of time that the metabolism needs to adjust itself. In short, all those crash diets and diets with their 'phases' of quick weight loss are not without risk in the long-term.

As has already been discussed in this book, primarily, a good diet must not focus on weight loss, but on the prevention of chronic ageing diseases. That is the essence of a 'real' diet. Moreover, the diet must be based on a thorough knowledge of human metabolism and in fact with the entire metabolic picture kept in mind. An item of food can indeed reduce the chance of cardiovascular disorders, but it can just as well increase the risk of dementia. In addition, the diet must fit into the evolutionary history of our species – our ancestors lived for hundreds of thousands of years without bread, pasta, potatoes, milk or large quantities of meat. Finally, the diet must also be based on insights originating from large and properly executed studies and not from small-scale studies, which are often cited by diet gurus to fit their purpose.

Some of the studies properly conducted are 'field studies'. A few well-known studies are those field studies researching population groups living in the Blue Zones. Blue Zones are areas in the world where people live longer on average and also stay in good health to a much greater age. Among such zones are the island of Okinawa on the coast of Japan, certain areas of Sardinia and a religious community in California (specifically, the Seventh-day Adventists who consider healthy nutrition as a pillar of their belief). In Okinawa, for example, there are five times more centenarians than in the West. This is not so much an issue of genes, because people from Okinawa who live in Brazil, die on average 17 years earlier than their family members in Okinawa.[282]

If we look at the common factors among these centenarians in the Blue Zones, then we see that they primarily consume a food pattern such as the one described in this book. They eat a lot of fruit and vegetables, legumes, nuts, fatty fish and herbs such as curcumin; they are sufficiently exposed to the sun (vitamin D); they drink little or no alcohol outside the occasional glass of red wine; they eat moderately (calorie restriction); they eat little or no red meat or dairy products; and they do (light) physical exercise on a regular basis.

However, Blue Zones do not always have to be on the other side of the world. Every person who eats a little healthily can create his or her own Blue Zone. A large study was published recently, in which 120,000 Dutch people were followed for 25 years. The men in the study who did not smoke, had a normal weight, exercised regularly and ate healthily (a lot of vegetables, fruit, nuts, legumes, fish and not too much meat) lived 8.5 years longer. The women who followed this healthy lifestyle, lived 15 years longer on average.[283] The investigators based their study on the Mediterranean diet, but as we have seen in this book, the diet described in this book can be even more healthy (and at least as delicious).

People are fascinated by longevity. We all wish to live to an old age and in good health. In this era of medicine and the human genome, it seems as if our health is mainly outside our control. Being healthy seems primarily an issue of good genes or the right medicines. However, medicine can solve little when it is too late and our genes primarily play a role in terms of maximum lifespan. For most people, lifespan is determined for 25 percent by genes, and for 75 percent by lifestyle. Only in people who get to 100 years or older, the genetic factor is much more important. So whoever wants to reach the age of 100 or older must have good genes. But whoever wants to solve a Sudoku puzzle when they are 80 or still wants to take a long trip in advanced old age should be interested in eating healthily. Moreover, the last decades of life will also be more productive, since a healthy nutrition pattern not only extends the lifespan, but the number of illness-free years will also increase exponentially.

This book set out to demonstrate the importance of food for our health. Somehow, it goes without saying that what we eat is crucial for our health. After all, every atom from which our body is constructed originates from the food that we eat. We are literally what we eat. Or, in the words of Walter de la Mare's poem:

> *It's a very odd thing –*
> *As odd as odd can be –*
> *That whatever Miss T. eats*
> *Turns into Miss T.*

Menus and recipes

It is unfortunate that in our society healthy food has become a lot more expensive than unhealthy food. Healthy food not only costs more money, it also costs more time in order to purchase and prepare. Putting healthy food on the table every day is not easy. Furthermore, fruit and vegetables should be bought continually because unlike biscuits or crisps, they spoil much faster. Also, unlike a candy bar, you cannot just take a piece of fruit out of the cabinet and eat it: you have to wash it or prepare it first.

It is unfortunate that healthy food and the important health benefits that come from it seem to have become more like an elite matter. That is why in the United Kingdom and the United States, many projects have been started to educate less affluent people in society about the importance of healthy food and how to convert this theory into practice.

However, this conversion of theory into practice is not always a straightforward matter. Here, a first initiative is provided to convert the theory of the food hourglass into practice. This chapter discusses various dishes and menus. These dishes are composed according to the food hourglass and are focussed on delaying the ageing process and the weight loss that automatically results from it.

Let me begin, however, by providing a brief description of what I eat myself during a weekday. That day begins as follows:

In the morning, just after getting up, I drink a glass of water (at room temperature, not from the refrigerator). Then

I make a cup of green tea with lemon. I eat breakfast 20 minutes later (first I do other various morning tasks such as washing, shaving, getting dressed, etc.).

During breakfast, I do the following:

- I dish out some oatmeal from a large bowl in the refrigerator (and that contains enough oatmeal for three days) and warm up the oatmeal in the microwave;
- I eat strawberries, one banana and some walnuts with the warm oatmeal;
- I eat purple grapes as dessert;
- I squeeze two oranges into a glass and add to it two spoonfuls of the remaining orange pulp;
- I swallow my iodine, magnesium, selenium and vitamin B nutrition supplements.

Around 11.00 in the morning, I begin to get a bit hungry and I eat a packet of dried fruit with nuts. Sometimes I eat an apple just before lunch so that my stomach is already full of fibre. This fibre delays the emptying of the stomach and levels off the sugar levels in my blood that will follow the lunch.

I have lunch at 13.00: salmon with salad, broccoli and mush-rooms. The dessert is a little container of soya yoghurt to which I add some blueberries.

I get hungry at 15.30. I eat an apple with a piece of dark chocolate. If I am really hungry, I eat oatmeal with a piece of dark chocolate.

I open a tin of mackerel at 18.30. I sprinkle some olive oil over it. I prepare a salad with olive oil, vinegar, pepper and curcumin. With this, I eat white beans seasoned with garlic, pepper, salt and homemade mayonnaise (based on olive oil).

I always prepare a bit extra, so I can take the leftovers to work the following day, in a water-tight bread bin.

I make sure that I have consumed two litres of liquids during the day, mainly in the form of water and two cups of green tea.

I eat nothing more between my dinner and the time that I go to sleep. I advise people who get hungry late in the evening, to eat something rich in fibre just before, after, or during their evening dinner, because fibre slows down the emptying of the stomach. This can be a bowl of oatmeal porridge, or a few teaspoons of fibre in a glass of water, a cup of soup, or soya dessert. Especially people who eat ample amounts of bread, pasta, potatoes and rice, tend to get hungry soon, sometimes already after one hour.

Breakfast

Oatmeal and other recipes
Oatmeal is consumed instead of bread or cereals, together with, for example:

- a piece of dark chocolate
- strawberries
- blueberries
- raspberries
- blackberries
- banana
- grapes
- dried fruit such as dried apricots or raisins
- grapefruit
- walnuts
- apple
- pear
- pomegranate
- linseed
- pumpkin seeds

The oatmeal:

- is made with vegetable milk such as soya milk (and not cow's milk). Ideally, one should use sugar free or very low sugar vegetable milk, which is available in many supermarkets;
- can be seasoned to taste with healthy sweeteners such as stevia or sugar alcohols (like erythritol);
- is best prepared in advance for convenience (in a portion for a few days) and kept in the refrigerator;
- can be warmed up in the microwave;
- can be seasoned to taste with cinnamon, stevia, cocoa powder, raisins, ground almonds, etc.

One of my favorite methods of preparing oatmeal porridge is to add oatmeal flakes to cooked soya milk, along with:

- cinnamon and chunks of apple *or*
- ginger cookie spices (cardamom, anise seeds, cloves, nutmeg), cinnamon and raisins.

Sometimes I melt dark chocolate in a saucepan, and cover my oatmeal porridge with a fat coating of chocolate, or I let the chocolate congeal in several deep plates and pour the oatmeal over it. Then I let the porridge become thicker in the fridge.

On the www.foodhourglass.com website you can find lots of other tasty oatmeal recipes.

If you want to eat something else for breakfast for a change, other than oatmeal porridge, these are some other alternatives to bread and breakfast cereals:

- soya yoghurt with nuts, fresh fruit (for example, sliced bananas), dried fruit and seeds like linseed or pumpkin seeds. Make sure the soya yoghurt is 'plain' and doesn't contain a lot of sugar;

- vegetables (for example, spinach or broccoli) with Quorn and beans;
- vegetables with tofu and beans;
- vegetables with an egg;
- a smoothie with fruit, nuts (for the fats) and a spoonful of vegetable protein powder (to ensure you ingest sufficient proteins in the morning). You can prepare a smoothie by mixing up all these ingredients in a blender;
- a smoothie with fruit, nuts, and tofu;
- different mashed fruits (a banana, apple, pear and kiwi), mixed with linseed and nuts;
- different nuts and blueberries in a bowl with almond milk.

Oatmeal with apricots and stevia

- oats
- 250ml soya milk (sweetened or unsweetened)
- 100ml orange juice (freshly squeezed)
- 4 dried apricots, cut into pieces
- a few drops of stevia

Pour the orange juice into a saucepan and bring it to the boil.

Add the pieces of apricot and let them plump up in the orange juice.

Pour the soya milk into a different saucepan, bring it to the boil and add the oats (use the amount of oats and cooking time as indicated on the packet) and add the stevia. Bring it to the boil again and stir well in the meantime. Allow the oatmeal to thicken.

Remove the apricots from the orange juice and add them to the oatmeal.

Allow the oatmeal to cool before serving.

Oatmeal with banana and cinnamon

– oats
– 250ml soya milk (sweetened or unsweetened)
– ½ banana, sliced
– cinnamon to taste
– salt to taste
– cocoa powder and grated coconut to taste

Pour the soya milk into a saucepan, bring it to the boil and add the oats (use the amount of oats as indicated on the packet) and pieces of banana. Allow the oatmeal to cook a few minutes until it has the desired consistency.

Add the cinnamon. Turn off the heat and let the oatmeal thicken for a few minutes.

Cocoa powder and grated coconut can be sprinkled on the oatmeal as garnish.

Fruit salad

– 300g grapes
– 1 orange, peeled and cut into pieces
– 1 apple, cored and chopped
– 1 pear, cored and chopped
– 1 tbsp agave nectar
– 2 tbsp almond slivers
– 100g soya yoghurt

Combine everything (note: do not peel the apple and pear, but rinse and dry them).

Tofu with leaf beetroot

– 250g tofu
– 1 cup leaf beetroot, finely chopped
– ¼ onion, finely chopped

- ¼ carrot, finely chopped
- 1 tbsp olive oil
- ½ tsp curry powder
- ½ tsp oregano
- ½ tsp basil (dried)

Rinse the tofu and cut it into little pieces. Fry the onion in a frying pan until slightly browned (this takes about 5 minutes).

Add the curry powder. Then add the tofu and the other ingredients.

Beverages with breakfast
It is healthy to consume the following beverages with breakfast:

- green tea, white tea or ginger tea
- 1 glass freshly squeezed fruit juice or vegetable juice (no store-bought juices)

To ensure that fruit juice doesn't cause such high sugar peaks (since fruit juice still always contains liquid sugars, apart from all the healthy stuff), it is best not to drink fruit juice on an empty stomach, but after you have eaten a meal, for instance, after breakfast. Always add the fibre that was left in the fruit juicer to your glass of juice, this will level off the sugar peaks. If you mix the fruit in a blender, the fibre will automatically be in the mixture. Also, you can prepare your fruit juice with added vegetables on a regular basis, since vegetables contain less sugar.

As we stated, it is better to make your own fruit juice and not buy it in the store (such juice hardly contains any fibre and often extra sugar is added). Although it is best to buy pomegranate juice in the store, because squashing pomegranates is difficult and you need a lot of them. Moreover, store-bought pomegranate juice is prepared with industrial machines, that squash the flesh but also the white pulp of the pomegranate. This pulp contains lots of healthy substances too. Make sure to buy 100 percent pomegranate juice, and not 30 percent pomegranate juice which has been watered down with water and sugar.

Don't always squash or blend (with a mixer) oranges, but vary as much as possible. Use lots of red and blue fruits as well (raspberries, strawberries, blueberries, bilberries, etc.). Although red and blue fruit tends to be expensive. You can solve this problem by buying frozen red and blue fruit in the supermarket. This is much cheaper and in this way you can also prepare your fruit juice in winter. Deep-frozen fruit still contains all the healthy nutrients.

This is a suggestion for a basic smoothie recipe: put a handful of deep-frozen raspberries, a banana and a pear into a blender. Mix it and you have made yourself an extremely healthy smoothie.

Numerous fruit and vegetable juice recipes are available on the internet.

Simplicity itself

– 2 oranges

Squeeze. Add 1 to 2 teaspoons of the pulp left behind in the press to the fruit juice.

Strawberry-grape juice

– a handful of strawberries
– a handful of blue grapes
– 1 banana

Blend everything together in a blender. Add a few walnuts, if you wish.

Strawberry-watermelon juice

- 150g strawberries
- 120g watermelon
- 1 teaspoon ground linseed
- a touch of cinnamon

Blend everything and add the cinnamon.

Beetroot-orange-apple juice

- 1 red beetroot
- 2 oranges
- 1 apple
- 1½ stalk of celery
- 1½ carrot

Wash and squeeze.

Vegetable juice with broccoli

- 1 carrot
- 1½ stalks of celery
- ¼ fennel bulb
- ½ red beetroot
- ½ stalk broccoli
- ¼ cucumber
- ¼ courgette

Wash and squeeze.

Cabbage-kiwi juice

- ½ cabbage or Savoy cabbage
- 1 beetroot
- 2 kiwis

Wash and squeeze.

Carrot-parsley juice

- 1 carrot
- 1½ stalks of celery
- ¼ cup parsley
- 2 spinach leaves
- ¼ red beetroot
- ¼ cup alfalfa

Wash and squeeze.

Lunch and dinner: make vegetables tastier

After a breakfast of oatmeal, fruit, nuts and dark chocolate there are two more main meals to be consumed. These meals are hot or cold meals with vegetables, meat or fish and, further, legumes and mushrooms as replacements for potatoes, pasta and rice.

The basis of these meals and the food hourglass is vegetables, with healthy highlights such as salad, broccoli, spinach and cabbage.

You don't always have to eat a single kind of vegetable with a meal. You can combine various types of vegetables, such as turnip-rooted celery with tomatoes, broccoli with swede, or fennel with carrots and broccoli.

What's important here is to elevate the TPC (Taste Per Calorie). In other words: make your vegetables taste better.

You will need to use a lot of herbs, vegetable broth, oils or onions in order to do this.

An example: you can season the broccoli with the swede by frying them in olive oil for 15 minutes, to which you have added red onion, garlic (powder), cayenne pepper and dill.

Vegetables, and especially salads, can also be seasoned to taste with:

- dressings (a sauce based on oil and various other ingredients) or vinaigrettes (a vinaigrette is a mixture of an oil with something sour, usually vinegar)
- nuts
- mushrooms
- tofu
- pieces of fish or poultry
- pieces of onion, shallots, radish, garlic
- onion powder and store-bought herb mixtures
- orange juice, apple juice, linseed (oil)
- pieces of apple, mandarin orange, pear
- peas and/or other legumes

A very simple salad dressing is 3 spoons of olive oil, 1 spoon of red wine vinegar and a bit of pepper and salt.

Vinaigrettes, dressings and mayonnaises

Lemon-basil vinaigrette

- ⅓ cup lemon juice
- 1 tsp basil
- ¼ cup olive oil (extra virgin)
- ¼ tsp garlic powder
- salt and pepper to taste

Mustard-garlic vinaigrette

- ½ cup balsamic vinegar
- ½ cup extra-virgin olive oil
- 1 clove finely chopped garlic
- 1 tsp mustard
- a pinch of salt
- black pepper to taste

Basil vinaigrette

- 1 cup olive oil
- ⅓ cup apple cider vinegar
- ¼ cup agave nectar (or regular honey)
- 3 tbsp basil freshly chopped
- 2 cloves garlic finely chopped

Broccoli dressing

- 1 stalk broccoli
- ¼ cup rice vinegar
- 2 tps Dijon mustard
- 2 cloves garlic chopped

Apple salad dressing

- 2 peeled apples
- ¼ cup freshly squeezed orange juice
- a pinch of cinnamon to taste

Stir all the ingredients together in a bowl.

Soya vinaigrette

- ½ cup balsamic vinegar
- ¼ cup agave nectar (or regular honey)

- ¼ cup extra-virgin olive oil
- 1 tsp soya sauce

Stir all the ingredients together in a bowl.

Mustard vinaigrette

- ⅓ cup balsamic vinegar
- ½ cup olive oil
- 2 tbsp Dijon mustard
- 1 tbsp agave nectar (or regular honey)
- a pinch of salt and pepper

Stir all the ingredients together in a bowl.

Homemade mayonnaise

- 250ml soya oil (or olive oil)
- 1 egg yolk
- 1½ tbsp herb vinegar
- ½ tsp black pepper
- ¼ tsp mustard powder
- a pinch of salt

Stir all the ingredients together in a bowl, except for the soya oil (or olive oil).

While continuing to stir, slowly add the soya oil (or olive oil) until the mixture begins to bind.

If the mayonnaise separates while stirring, then add one or two teaspoons of warm water.

Olive oil mayonnaise

- 250ml olive oil
- 2 egg yolks
- 2 tbsp lemon juice
- ¼ tsp salt

Continued overleaf

Mix the egg yolks and the salt in a bowl with a whisk until the mixture is light yellow and foamy.

While continuing to whisk, slowly add the olive oil until the mixture begins to bind. Then you can add the oil a bit faster.

Halfway through, slowly add the lemon juice.

Add salt to taste.

Tuna mayonnaise

- 100g tuna
- 3 anchovy fillets
- 50ml olive oil
- 1½ tbsp lemon juice
- 3 tbsp capers
- 300ml olive oil mayonnaise (see above)
- 4 black olives, pitted
- 1 lemon, cut in pieces

Use a food processor or hand-held blender to purée the tuna, anchovies, olive oil, lemon juice and half of the capers into a sauce.

Mix the tuna purée with the olive oil mayonnaise.

Season to taste with salt.

Lunch and dinner: examples

A healthy main meal consists of fatty fish, poultry, tofu or Quorn with vegetables, mushrooms and legumes as replacements for potatoes, pasta or rice.

It is best to eat two such meals a day, for example, a larger warm meal at noon, and a lighter cold meal in the evening.

To save time you can prepare extra portions on some days. You can store these extra portions of vegetables or legumes in storage tins, in your fridge, and use them in the next couple of days. You can also save the leftovers of previous meals in such tins. This way, you won't need to prepare all your vegetables or legumes twice a day.

Another alternative for potatoes, pasta or rice is quinoa. Quinoa looks a bit like round rice. But it isn't a grain, it's related to spinach. Quinoa contains a lot of vegetable amino acids and has a relatively low glycaemic index.

If you still have to choose between potatoes, pasta or rice, it's best to choose rice. Brown or black rice, of course, since white rice causes high sugar peaks. The advantage of rice is that it doesn't contain immunogenic proteins such as gluten, which can cause allergies or intolerances in some people. Gluten is found in pasta and grains of which bread is made such as wheat, barley, or rye, but not in rice, and very little in oatmeal.

If you decide to eat bread after all, then use wholemeal bread, preferably rye bread. You can cover this wholemeal bread with cheese, white meat such as poultry, or with healthy alternatives to meat products, such as pesto, hummus or vegetable paté. Pesto is made from olive oil, basil, nuts and cheese. Hummus is a kind of paté made from chickpeas. Both are available in the supermarket. You can also add olives, capers or rocket to the pesto or hummus.

Portobello, chickpeas and tomato

- 2 Portobello mushrooms, thinly sliced
- 1 tin chickpeas (about 400g); drained, but reserve the liquid
- 1 tomato, cut into cubes
- 1 onion, chopped
- ½ teaspoon olive oil
- ½ cup red wine
- 2 cloves garlic, finely chopped

Heat the olive oil gently in a saucepan.
Stir-fry the garlic and onion for a couple of minutes.
Add the mushrooms and red wine.
Fry for about five minutes.
Add the tomatoes, chickpeas and half of the reserved chickpea liquid.
Fry for another 5 to 10 minutes.
Serve.

Stuffed marrow with minced chicken

- 300g minced chicken
- 1 marrow
- 2 tomatoes
- 1 red pepper, chopped
- 1 green pepper, chopped
- 1 onion chopped
- 1 egg beaten
- 1 clove garlic, pressed
- ½dl bouillon
- salt and black pepper

Preheat the oven to 170 degrees.
Peel the marrow and cut it into four pieces. Cut the pieces

length-wise and scoop out the pulp. Cut the pulp into pieces.

Mix the pulp with the mince, the onion, the garlic and the egg. Add pepper and salt to taste.

Fill the hollowed-out pieces of marrow with the mince. Add the tomatoes, peppers and bouillon.

Place the dish in the oven for 30 minutes.

Serve with salad.

Salad with mushrooms, turkey and tofu

- 1 head lettuce
- 250g tofu
- 200g turkey fillet, cut in strips
- 250g shiitake mushrooms, cut in strips
- 100g spring onions, cut in rings
- 80g linseed
- 1 tbsp ginger, chopped
- 3 tbsp olive oil (or a different vegetable oil)
- 3 tbsp rice vinegar
- 3 tbsp soya sauce
- salt and black pepper

Fry the tofu in olive oil in a frying pan. Then cut the tofu into cubes.

Fry the meat in the remaining olive oil. Season the meat with salt, black pepper and ginger.

Add the shiitake mushrooms and spring onions.

Add the linseed and mix everything.

Mix the rice vinegar and soya sauce together and then sprinkle the mixture over the dish.

Mackerel fillets with onion, shallots and garlic

– 500g mackerel fillets
– 1 onion, finely sliced
– 3 shallots, finely sliced
– 1½ cloves garlic, pressed
– ½ tbsp vinegar
– olive oil
– pepper and salt

Fry the mackerel fillets in olive oil until done. Then place the fillets on a warmed dish.

Sauté the shallots, onions and the garlic until they are lightly browned. Add the vinegar.

Add this mixture to the mackerel fillets.

Serve with vegetables such as broccoli or salad.

Tomatoes with goat cheese

– 400g tomatoes, cut in slices
– 4 round slices goat cheese
– 200g black olives, cut in half
– 4 tbsp hazelnut oil
– ½ bunch thyme, finely chopped
– ½ bunch oregano, finely chopped
– salt and black pepper to taste
– olive oil (or soya oil)

Preheat the oven to 180 degrees.

Brush the bottom of a baking dish with the olive oil.

Place the tomato slices in the dish and sprinkle them with pepper and salt.

Place the olives and the goat cheese on top of the tomato slices and pour the hazelnut oil on top.

Sprinkle everything with the thyme and oregano.

Let the dish cook au gratin in the oven for four minutes.

Salmon fillet with vegetables and shrimp (for four people)

- 4 salmon fillets
- 300g shrimp
- 2 courgettes, cut in long, thin slices (less than 0.5 cm thick)
- 3 eggs
- 2 tbsp lemon juice
- 8 tbsp olive oil
- 20ml anise liqueur
- 1 splash mineral water
- salt and black pepper

Mix the lemon juice with three tablespoons olive oil, salt and pepper and brush the mixture on the salmon fillets.

Wrap the courgette slices around the salmon fillets.

Heat a few tablespoons of olive oil in a frying pan and fry the salmon fillets for about five minutes.

Sprinkle the shrimp with anise liqueur.

Beat the eggs with a splash of mineral water, salt and pepper.

Add the shrimp to the egg mixture and fry the mixture in a frying pan with olive oil for three minutes.

Serve.

Steamed leeks with mackerel

- 500 mg mackerel fillets
- 1 onion, finely chopped
- 1 leek, finely chopped
- 1 sprig dill
- 1 sprig parsley
- ½ tbsp vinegar
- 5 tbsp olive oil
- salt and black pepper

Continued overleaf

Sprinkle the mackerel fillets with salt and a little black pepper.

Place the mackerel in a baking dish brushed with olive oil. Add the onion and leek, together with the dill and parsley. Mix the oil with the vinegar and pour it over the fillets.

Cover the dish with aluminium foil and let the fish cook in the oven at 190 degrees.

Remove the sprigs of parsley and dill and add fresh sprigs before serving.

Tofu/soyabeans with spinach and tomatoes

- 1 block tofu (about 400g). The tofu can be replaced by beans (of any kind)
- 1 box frozen spinach (about 300g), defrosted
- 3 tomatoes, cut in pieces
- 2 tbsp lemon juice
- ⅛ tsp cayenne (red pepper)
- ⅛ tsp onion powder
- ½ cup vegetable bouillon

Stir-fry all the ingredients in the vegetable bouillon. Serve.

Chicken salad with spinach, avocado and olives

- lettuce
- 2 heads fresh spinach, cut in pieces
- 1 cup roast chicken, cut in pieces
- 1 avocado, peeled and cut in pieces
- 1 cup soaked oats (first allow to soak in a bowl of boiled water for 30 minutes)
- ¼ cup spring onion, cut in pieces
- ¼ cup fresh parsley, chopped
- 12 olives, pit removed and cut into four pieces

Mix everything.

The lemon-basil vinaigrette can be added to the mixture (see vinaigrettes).

Serve.

Steamed vegetables with chicken breast

- 500g tomatoes, peeled and cut into cubes
- 200g aubergines, sliced
- 200g courgettes, sliced
- 200g red pepper, blocks
- 150g red onion, cut in slices
- 4 tsp salt
- 2 tbsp olive oil
- 2 cloves garlic, finely chopped
- 10 basil leaves, chopped
- 3 sprigs thyme
- 1 bay leaf
- salt and black pepper
- chicken breasts (already prepared)

Place the slices of courgette and aubergine in a sieve. Sprinkle them with the salt and let the liquid be extracted for 30 minutes.

Afterwards, rinse the aubergine and courgette slices under cold water and pat them dry.

Heat olive oil in a frying pan and fry the onion in it over medium heat.

Add the aubergine and courgette, together with the red pepper and garlic. Fry for a few minutes until the pepper is soft.

Add the herbs and tomatoes. Season to taste with pepper and salt.

Let simmer for 30 minutes.

Serve the vegetables with fried chicken breast.

Spinach with goat cheese

- 500g spinach
- 1 piece of goat cheese with rind
- vinegar

Cook the spinach briefly in a small amount of water. Drain the spinach and slice it.

Place the spinach in a baking dish, add vinegar and some salt and place the piece of goat cheese on top.

Place the dish in the oven and wait until the cheese begins to melt a little and turns light brown.

Serve.

Salad with smoked salmon and broccoli

- 5 slices of smoked salmon, cut in strips
- 500g peas
- 100g broccoli, in small rosettes
- 100g green beans, the ends removed
- ½ cup soya yoghurt
- 1 tbsp olive oil
- 1 lemon, squeezed
- 1 bunch dill, finely chopped
- salt and black pepper

Cook all the vegetables, one by one, in a large pot of water, for three to five minutes until they are al dente.

Mix the olive oil, lemon juice, dill and soya yoghurt in a bowl, and season the mixture to taste with pepper and salt.

Add the vegetables and salmon.

Serve.

Ragout of asparagus, mushrooms and quail eggs

- 11 quail eggs
- 500g mushrooms, sliced
- 1½kg green asparagus, cut in pieces (about 3 cm)
- 1 litre vegetable stock
- 3 tbsp olive oil (or a different vegetable oil)
- 1 bunch parsley, finely chopped
- salt and black pepper

Boil the quail eggs until hard (boil for eight minutes and then run them under cold water).

Heat the vegetable stock and cook the asparagus in it (about 12 minutes).

Fry the mushrooms for about four minutes in a frying pan with olive oil.

Drain the asparagus and add the mushrooms.

Add eight tablespoons of the asparagus cooking liquid to the mixture.

Add the parsley. Season to taste with salt and pepper.

Peel the quail eggs, cut them in quarters and add them to the dish.

Fennel salad with Roquefort

- lettuce
- 200g Roquefort, cut into cubes
- 2 fennel bulbs, cut in slices
- 100g raisins
- 100g walnuts
- 1 stick of celery, finely sliced
- 1 tbsp tapenade (this is a mixture of olives, capers, anchovy and olive oil)
- 1 lemon
- 4 tbsp olive oil
- salt and black pepper

Continued overleaf

Let the raisins plump in lukewarm water for 10 minutes.

Put the fennel and celery in a bowl and add the lemon juice, olive oil and tapenade. Season to taste with pepper and salt and allow to marinate for 20 minutes.

Add everything together and serve.

Stuffed pepper with salmon

- 160g smoked salmon, cut in small pieces
- 2 yellow peppers
- 150g cream cheese
- ¼ cucumber, peeled and cut in cubes
- 4 tbsp homemade mayonnaise (see above)
- 1 tbsp dill
- salt and pepper

Mix the mayonnaise with the cream cheese in a bowl.

Add the cucumber, dill and salmon, together with some pepper and salt.

Cut the peppers in half and remove the seeds. Fill the peppers with the salmon.

Garnish with sprigs of dill and slices of lemon.

Serve together with salad.

Turbot fillet with red lentils and broccoli

- 4 turbot fillets (120g per piece)
- 200g tinned red lentils
- 100g broccoli
- 1 shallot, chopped
- 4 tbsp olive oil
- 1 tsp turmeric powder
- 3 cloves garlic, finely chopped
- 25g chives, finely chopped

- 500ml chicken bouillon
- salt and black pepper

Pour two tablespoons of olive oil into a saucepan. Sauté the lentils, broccoli, shallot, garlic and turmeric powder over medium heat for three minutes.

Add the bouillon and gently simmer for four minutes. Add salt, pepper and chives. Keep this mixture warm.

Make a few gashes in the turbot fillets.

Pour two tablespoons of olive oil into a frying pan and fry the fillets for four minutes. Add salt and pepper.

Serve on a bed of lentils and broccoli.

Avocado, tofu and turkey

- 2 avocados, of which the flesh has been cut into cubes
- 150g tofu, cut into cubes, sprinkled with salt and pepper
- 200g turkey breast, cut in pieces
- 1 red pepper, cut in pieces
- 100g alfalfa or radish sprouts
- 2 tbsp olive oil
- 2 tbsp soya sauce
- lemon juice, salt and black pepper to taste

Mix the cubes of avocado, tofu, turkey and the alfalfa/radish sprouts in a bowl.

Stir-fry everything in olive oil and add the soya sauce to taste.

Serve.

Caesar salad

- 4 anchovy fillets
- 3 eggs
- 3 cloves garlic, finely chopped
- 1 lemon, squeezed
- 80g grated Parmesan cheese
- 3 tbsp olive oil
- 1 bunch tarragon, chopped
- 1 bunch chives, snipped
- a few walnuts, broken in half
- lettuce

Tear the lettuce in strips.

Boil two eggs until hard (eight minutes) and run them under cold water before peeling them.

Mix the garlic, lemon juice, one raw egg and the olive oil in a bowl. Add the lettuce, chives and tarragon.

Place everything on a plate and sprinkle the Parmesan and walnut pieces on top.

Salmon with green cabbage, cumin seed and garlic

- 500g green cabbage, sliced
- 1 piece salmon, 300 grams
- ½ tbsp cumin seed
- 1½ tbsp olive oil
- 80ml tomato sauce
- salt and black pepper
- 1 clove garlic, finely chopped

Cook the slices of green cabbage in water for one minute.

Cool them off under cold running water.

Pour olive oil into a large pan.

Fry the cabbage with the cumin seed for two minutes over high heat. Add salt and pepper.

Cut the salmon in half. Sprinkle the pieces with salt and pepper and steam them until done.

Heat up the tomato sauce and add the garlic.

Place the salmon fillets on a bed of cabbage and add the tomato sauce.

Desserts

Some examples of healthy and simple desserts:

- a bowl of strawberries, raspberries or blueberries with a piece of dark chocolate;
- a bunch of grapes;
- dried fruit with nuts;
- an apple, pear or banana.

Berry dessert

- 75g blueberries, blackberries or raspberries
- 2 tsp (soya) lecithin
- walnut pieces
- 1 tsp lemon juice to taste

Chocolate mousse (for six persons)

- 400g dark chocolate (at least 70 percent cacao)
- zest from 1 orange
- 8 eggs
- 4 tsp instant coffee
- ½ glass rum (3½cl)
- 1 pinch salt

Chop the chocolate in pieces and put them in a saucepan.

Add ½ cup very strong coffee and the rum to the saucepan.

Continued overleaf

Melt the chocolate mixture over low heat (add a little water if the mixture is too thick).

Add half of the orange zest to the saucepan.

Break the eggs, separating the yolks into one bowl and the whites into a different bowl. Add a pinch of salt to the egg whites and beat the egg whites until stiff.

Pour the melted chocolate mixture into the bowl with the egg yolks.

Mix well and allow to cool a bit.

Pour the melted chocolate mixture into the bowl with the egg whites and mix well.

Sprinkle the mixture with the other half of the orange zest.

Place the mousse in the refrigerator and allow to chill for a few hours.

Apple cake

- ½ cup oats
- 3 apples, peeled and cut in pieces
- ¼ cup apple juice
- ¼ cup vanilla soya milk
- ¼ tsp vanilla
- 1 egg white
- ¼ cup raisins, chopped
- 1 tsp cinnamon

Mix the vanilla with the apple juice.

Add the soya milk, egg white and cinnamon.

Add the apples, raisins and oats.

Bake the mixture (uncovered) at 170 degrees for one hour.

Cover with aluminium foil to store.

Soya yoghurt shake with fruit

- 100g strawberries
- 100g raspberries
- 70g cranberries
- 70g blueberries
- 1 banana
- 70g agave nectar (or honey)
- 300g soya yoghurt

Mix all the ingredients with a mixer.

Orange sorbet

- 1 litre orange juice (freshly squeezed)
- 2 tbsp orange liqueur (such as Cointreau)
- ¼ tsp lemon juice

Mix everything and freeze in ice cube trays.
    Take the trays out of the freezer (and wait a few minutes until the cubes slowly begin to melt).
    Remove the cubes from the tray and mix or blend them.
    Serve immediately, with pieces of strawberry, if desired.

Tofu pudding with dark chocolate and blueberries

- 350g silken almond tofu (available in some natural food shops or via internet)
- 300g dark chocolate (more than 70 percent cacao)
- blueberries (or strawberries, raspberries or blackberries)

Blend the tofu in a mixer for 30 seconds until creamy.
    Melt the chocolate in a saucepan.
    Add the melted chocolate to the tofu.
    Beat the mixture for one minute with a mixer.
    Garnish with blueberries and serve.

Peach mix

- 2 dates
- 3 peaches (or nectarines)
- 1 frozen banana
- ¼ cup vanilla soya milk
- 1 tsp vanilla extract
- ⅛ tsp cinnamon

Cut the fruit into pieces.
Mix all the ingredients in a mixer (blender).

For health professionals

Health professionals who want to find out more about the scientific background of the food hourglass can read the short scientific article below. This article has been written with an emphasis on type 2 diabetes. However, the food hourglass is aimed at reducing the risk of many aging-related diseases, among them type 2 diabetes.

The food hourglass: a novel concept for nutrition guidance in type 2 diabetes & weight loss

Problem

Currently, patients diagnosed with type 2 diabetes are advised to follow dietary guidelines issued by national organisations, like the American Diabetes Association (ADA) or the National Institute for Health and Care Excellence (NICE). However, the health benefits that are brought about by these recommendations are rather modest. For example, patients following the American Diabetes Association guidelines reduce on average their HbA1c with 0.4 percent.[i] Yet there exist various diets that could have a more substantial health impact. For example, a more vegetarian-oriented diet for diabetes patients is three times as effective in reducing HbA1c1 compared with patients following the ADA diet guidelines.[i] A low glycaemic index diet made it 75 percent less likely to increase diabetes medication than when following an ADA diet[ii]. A hypocaloric mainly vegetable-based diet can even reverse diabetes in all subjects in eight weeks time.[iii] These and similar studies show that nutrition can substantially impact on diabetes. However, providing diet guidelines that can be adhered to by patients in the long term and that can change often deeply rooted eating habits are not easy to implement.

Solution

We developed the food hourglass, a nutrition model to help type 2 diabetes patients to make healthier food choices, in order to improve their metabolic and cardiovascular health and lose weight. The food hourglass is an alternative to a low-fat or low-carb/high-protein diet for diabetes patients. With the food hourglass we want to introduce a new type of diet, called a low glycaemic load *healthy macronutrient diet* (HMD).

The food hourglass is a figure in the shape of an hourglass (2 triangles facing each other). The upper triangle contains less healthy food of which the intake should be reduced, while the lower triangle contains healthier food of which the intake should be increased.

Each coloured layer in one triangle has a corresponding coloured layer in the opposite triangle so that patients can clearly see how to replace one food by another healthier alternative. For example, red meat in the red layer in the upper triangle can be replaced with poultry or fish in the corresponding red layer in the lower triangle.

The food hourglass draws on insights from diverse fields like endocrinology, biogerontology, evolutionary medicine and nutrition science. The food hourglass emphasises a substantial reduction of specific starchy foods (temporarily or indefinitely). These starchy foods (SFDs) are bread, potatoes, pasta and rice. Comparable recommendations can be found in the Harvard food pyramid, the Mayo Clinic food pyramid and the official Austrian and Swiss food pyramid. These food models put less emphasis on starchy foods like (whole grain) bread, potatoes, pasta or rice. For example, the base of the Mayo Clinic Food Pyramid and Austrian food pyramid (originally composed of bread, potatoes, pasta and rice) is completely replaced by vegetables and fruits. Potatoes have been moved to the off limit top of the Harvard food pyramid (together with red meat), in the same category as soda and sweets.

The rationale behind these changes is that high glycemic index (GI) and high glycemic load (GL) diets are comprised of foods that produce high blood sugar peaks (high GI) and deliver large amounts of easy-digestible carbohydrates in the form of sugars and starch (high GL), which increases the risk of metabolic diseases like cardiovascular disease and type 2 diabetes. For example, a prospective study following 15,714 women for nine years showed that the highest quartile of dietary glycaemic load was associated with a 47 percent increased risk of cardiovascular disease compared to the lowest quartile, further increasing to 78 percent for overweight women.[iv] A high glycaemic load and low fibre diet was associated with a doubling of the risk of type 2 diabetes (N = 44 000; RR 2,17).[v] A prospective study following a cohort of 75,521 women for ten years showed that the risk of coronary heart disease for the highest glycaemic load quintile was 98 percent higher.[vi] A prospective study of 64,227 Chinese women for 4 to 6 years showed a correlation between glycaemic index, glycaemic load and type 2 diabetes, especially for the glycaemic load and rice intake.[vii]

Additionally, studies have shown that low glycaemic index diets are superior to low-fat diets in terms of improving cardiovascular parameters and weight loss. A Cochrane meta-review showed that ad libitum low glycaemic index diets are superior to caloric restricted low-fat diets in terms of lipid profiles and weight loss.[viii] Adhering to a low glycaemic index diet improves metabolic parameters more than an isocaloric low-fat diet (challenging the assumption that a calorie is always a calorie).[ix] A hypocaloric mainly vegetable-based diet without bread, potatoes, pasta and rice can reverse diabetes in eight weeks time.[iii]

In the food hourglass, we do promote the use of one particular grain product, namely oatmeal, which can serve as a substitute for bread, especially during breakfast. Oatmeal has a lower glycaemic index and load than white and brown bread (and most wholemeal breads), and contains oat fibre (water soluble beta-glucans) that have a beneficial impact on cholesterol and glucose levels. In contrast to wholemeal bread, oatmeal has been granted a health claim by the European Food Safety Authority (EFSA) and can confer benefits to diabetes patients. Diabetes patients with difficult to regulate glucose levels who were put on a two-day oatmeal diet, were able to reduce their insulin medication by 40 percent, an effect that lasted for at least four weeks after this intervention.[x] Consuming oatmeal instead of higher glycaemic index cereals and bread, resulted in lower blood sugar levels in type 2 diabetes patients.[xi]

The food hourglass recommends patients to replace bread and cereals with oatmeal during breakfast, and to replace potatoes, pasta and rice with legumes (beans, lentils, peas, . . .), mushrooms and an extra portion of (other) vegetables for lunch and dinner. For example, a study among 1,879 adults showed that substituting one serving of beans for one serving of white rice was associated with a 35 percent lower risk of the metabolic syndrome[xii]. The food hourglass advocates vegetables, legumes, oatmeal and fruit as basic bulk foods to improve blood sugar levels and to attain healthy weight loss.

Besides the reduction of starchy high glycaemic load

foods, the food hourglass recommends replacing red meat more often with white meat and fish. This is because an increased intake of red meat is associated with an increased risk of cardiovascular disease, diabetes and cancer[xiii] (N= 120 000). A 12-year follow-up study of 27,147 individuals showed a direct correlation between meat intake, total protein intake and type 2 diabetes.[xiv] A study with almost half a million Europeans showed a 44 percent increase in mortality in people who ate more than 160 grams of processed meat a day compared to 10-19,9 grams per day.[xv] Substituting red meat with healthier protein sources like chicken reduces mortality by 14 percent.[xiii] Increasing the intake of fish, especially fatty fish, is associated with reduced cardiovascular mortality and improved cardiovascular parameters.[xvi] Red meat can also be replaced with tofu (soy) and quorn (a protein-rich food product made of a fungus).

Regarding dairy, cheese is the only dairy product that is recommended, being an important source of menaquinone (vitamin K2) and trans-palmitoleic acid in the western diet. Milk is not recommended because of the insulinotropic effects of milk[xvii, xviii] and other possible negative health effects of milk in the long term.[xix, xx, xxi] Soda and other sugar-sweetened beverages are adviced against (like commercial low-fibre fruit juice). Ample amounts of water, and further tea, alcohol in moderation, low-sugar vegetable milk and coffee are healthier drink alternatives. The food hourglass further advocates healthy snacks (like nuts, dark chocolate, soy yoghurt, fruit, . . .) and the use of healthier sugar substitutes (like stevia, tagatose, sugar alcohols and other compounds that result in reduced sugar and insulin increments) and healthy vegetable fats.

Further discussion

It's well known that it's often difficult for patients to adhere to dietary guidelines, especially in the long term. The food hourglass can make long-term compliance more likely and could provide more pronounced health benefits for type 2 diabetes patients compared to low-fat diets and low carb/high-protein diets.

Firstly, a model shaped like a food hourglass can make a more clear distinction between food products that are recommended (in the lower triangle) and food products that should be consumed less (in the upper triangle). Additionally, the corresponding similar coloured layers in both triangles clearly provide alternatives for the dissuaded foods.

While we do put limits on portions for certain foods (like alcohol, chocolate or meat) we want to keep the dietary guidelines as clear and simple as possible. We deliberately don't want to encumber patients with calorie counting, calculating protein or fat content or weighing foods, methods that can reduce compliance in the long term. As studies show, low glycaemic index and glycaemic load diets negate the need for calorie-restricted diets: patients can follow ad libitum low GI/GL diets and still lose weight and improve cardiovascular parameters.[viii]

The food hourglass model enables patients to quickly see and understand some important dietary principles (less high glycaemic load foods, more healthy protein sources, more healthy fats), while at the same time providing clear alternatives that can direct them to healthier food choices. In this way, the food hourglass can also be very useful for medical practitioners in a clinical setting, when they want to provide extensive dietary advice but only have a limited amount of consultation time.

Regarding the content of the food hourglass, the model

not only incorporates the hypothesis that an increased intake of unhealthy fats increases the risk of cardiovascular disease and weight gain, but also includes the more recent hypothesis that high glycaemic load foods play a role in metabolic diseases and the obesity epidemic as well.[xxii] Contrary to low-carb/high-protein or low-fat diets, the food hourglass doesn't advise against or favours specific macronutrients (carbs, fats or protein). Instead, it recommends the use of more healthier foods in each macronutrient group, namely low glycaemic index/load carbohydrates, healthy protein sources (fish, poultry, vegetable protein, . . .) and healthy fats. We call this diet a low glycaemic load *healthy macronutrient diet* (HMD), that can serve as an alternative to low-carb high-protein and low-fat diets.

In conclusion, the food hourglass wants to empower patients to make healthier food choices by providing a model that quickly and clearly explains some important dietary principles that aim to improve metabolic parameters, reduce the risk of ageing-associated diseases and attain weight loss.

i. A low-fat vegan diet improves glycemic control and cardiovascular risk factors in a randomized clinical trial in individuals with type 2 diabetes. *Diabetes Care* 29, 1777–83 (2006).
ii. A randomized clinical trial comparing low-glycemic index versus ADA dietary education among individuals with type 2 diabetes. *Nutrition* 24, 45–56 (2008).
iii. Reversal of type 2 diabetes: normalisation of beta cell function in association with decreased pancreas and liver triacylglycerol. *Diabetologia* 54, 2506–14 (2011).
iv. High dietary glycemic load and glycemic index increase risk of cardiovascular disease among middle-aged women: a population-based follow-up study. *J. Am. Coll. Cardiol.* 50, 14–21 (2007).
v. Dietary fiber, glycemic load, and risk of NIDDM in men. *Diabetes Care* 20, 545–50 (1997).
vi. A prospective study of dietary glycemic load, carbohydrate intake, and risk of coronary heart disease in US women. *Am. J. Clin. Nutr.* 71, 1455–61 (2000).
vii. Prospective study of dietary carbohydrates, glycemic index,

glycemic load, and incidence of type 2 diabetes mellitus in middle-aged Chinese women. *Arch. Intern. Med.* 167, 2310–6 (2007).

viii. Low glycaemic index or low glycaemic load diets for overweight and obesity. *Cochrane Database Syst. Rev.* CD005105 (2007). doi:10.1002/14651858.CD005105.pub2

ix. Effects of dietary composition on energy expenditure during weight-loss maintenance. *JAMA* 307, 2627–34 (2012).

x. Clinical benefit of a short term dietary oatmeal intervention in patients with type 2 diabetes and severe insulin resistance: a pilot study. *Exp. Clin. Endocrinol. Diabetes* 116, 132–4 (2008).

xi. Beneficial effects of high dietary fiber intake in patients with type 2 diabetes mellitus. *N. Engl. J. Med.* 342, 1392–8 (2000).

xii. A higher ratio of beans to white rice is associated with lower cardiometabolic risk factors in Costa Rican adults. *Am. J. Clin. Nutr.* 94, 869–76 (2011).

xiii. Red meat consumption and mortality: results from 2 prospective cohort studies. *Arch. Intern. Med.* 172, 555–63 (2012).

xiv. High intakes of protein and processed meat associate with increased incidence of type 2 diabetes. *Br. J. Nutr.* 109, 1143–53 (2013).

xv. Meat consumption and mortality – results from the European Prospective Investigation into Cancer and Nutrition. *BMC Med.* 11, 63 (2013).

xvi. Fish and long-chain omega-3 fatty acid intake and risk of coronary heart disease and total mortality in diabetic women. *Circulation* 107, 1852–7 (2003).

xvii. Glycemia and insulinemia in healthy subjects after lactose-equivalent meals of milk and other food proteins: the role of plasma amino acids and incretins. *Am. J. Clin. Nutr.* 80, 1246–53 (2004).

xviii. Milk consumption and circulating insulin-like growth factor-I level: a systematic literature review. *Int. J. Food Sci. Nutr.* 60 Suppl 7, 330–40 (2009).

xix. Animal foods, protein, calcium and prostate cancer risk: the European Prospective Investigation into Cancer and Nutrition. *Br. J. Cancer* 98, 1574–81 (2008).

xx. Dairy products, calcium, and prostate cancer risk in the Physicians' Health Study. *Am. J. Clin. Nutr.* 74, 549–54 (2001).

xxi. Consumption of milk and calcium in midlife and the future risk of Parkinson disease. *Neurology* 64, 1047–51 (2005).

xxii. Diet and cardiovascular disease prevention the need for a paradigm shift. *J. Am. Coll. Cardiol.* 50, 22–4 (2007).

Glossary

Adenosine triphosphate (adenosine-tri-'phosphate', ATP): these are the molecules in the body that keep everything running. ATP molecules react with proteins by adhering to them. As a result, those proteins change structure so that they can perform a specific function. An ATP molecule can, for example, adhere to a channel protein in the cell wall, whereby it opens and certain molecules can enter the cell.

Alzheimer's disease: see *dementia*.

Alzheimer's: see *dementia*.

Amino acid: amino acids are the building blocks of proteins. An amino acid is a small molecule constructed out of always the same set (or 'backbone') of nine atoms with a further few specific atoms that determine the type of amino acid. In the body, there are 20 types of amino acids that attach to each other in chains that form proteins. One protein consists of tens to thousands of amino acids (see also *protein*).

Antioxidant: a substance that neutralises free radicals because it reacts easily with them. See also *free radical*.

Arteriosclerosis: See *atherosclerosis*.

Atherosclerosis: is also called arteriosclerosis. This is the clogging of the arteries. These are, among others, the arteries near the heart or the brain. When an artery suddenly completely clogs in the heart, then part of the heart dies off due to lack of oxygen and physicians call that a heart attack. When an artery completely clogs in the brain, then that is a stroke.

Atom: the building block of all material. An atom consists of an atomic nucleus and electrons that whirl around the nucleus. The protons and neutrons are located in the atomic nucleus. The number of protons in the nucleus determines the name of the atom. Hydrogen has one proton, iron has 26 and uranium has 92.

ATP: see *adenosine triphosphate*.

Bacteria: a one-cell organism that has no cell nucleus.

Base or base molecule: is a molecule such as guanine (G), cytosine (C), adenine (A) or thymine (T). A base always consists of one group of around 15 atoms of carbon, nitrogen, hydrogen and oxygen. Two bases always form the rungs from which the DNA ladder is constructed.

Billion: thousand million (a 1 with 9 zeroes).

Brain cell: see *neuron*.

Cancer: see *mutation*.

Carbohydrate: sugar. A carbohydrate can consist of one sugar molecule (such as glucose), two sugar molecules (such as granulated (table) sugar, which is constructed out of glucose and fructose), or many thousands of sugar molecules (such as starch, which is constructed out of thousands of glucose molecules). See also *glucose*.

Cardiovascular disease: the slow process whereby blood vessels clog up. When a blood vessel suddenly clogs completely in the heart, then it is a heart attack. See also *heart attack* or *stroke*.

Cell fluid: this is the space in a cell outside of the cell nucleus. The fluid in our cells consists of water in which proteins, mitochondria and other cell components float.

Cell membrane: the cell wall. The cell membrane consists of fatty molecules that encapsulate the water and proteins that make up a cell.

Cell nucleus: cells contain a cell nucleus at their core, in which the DNA is located.

Cellular skeleton: see *cytoskeleton*.

Cerebral cortex (or brain cortex): is the outer part of the brain. The cortex is a layer, a few millimetres thick, that forms the surface of the cerebrum. The cortex consists of oblong pyramid cells that are arranged in layers and it is here that consciousness occurs. When surgeons send some electrical current through the cortex, then it seems as if we have been touched somewhere or a long forgotten memory is suddenly recalled.

Cortex: see *cerebral cortex*.

Cortisol: a hormone originating in the adrenal gland that is released upon stress. See also *hormone*.

Crosslink: a connection between two proteins. This connection can be constructed from one sugar molecule. Crosslinks between the collagen fibres in the skin can make the skin more 'rigid' and less elastic as a result of which the skin becomes wrinkled.

Cytoplasma: see *cell fluid*.

Cytoskeleton: long interlocking tubes constructed from proteins. Small molecules like ATP (see adenosine triphosphate) attach themselves to the cytoskeleton, as a result of which its structure changes, as does the entire cell. Because of this, cells can propel themselves, attach themselves to other cells or eat bacteria.

Dementia: the dying away of brain cells, usually due to proteins that clump together in and around the brain cells and in so doing suffocate the cells. Depending on the type of proteins and the areas of the brain that are the most affected, there is Alzheimer's, Lewy body dementia or frontotemporal dementia. One form of dementia arises due to numerous mini-strokes in the brain (caused by, for example, high blood pressure or diabetes). This is called 'vascular dementia'. See also *stroke*.

Diabetes: a disease where too much sugar remains circulating in the bloodstream. In type-2 diabetes, this is caused because the liver, fat and muscle cells no longer react sufficiently to insulin, which is a substance that ensures cells can absorb sugar. Because of this, the sugar remains circulating too long in the bloodstream and other body cells, such as the kidney, eye or nerve cells, become damaged by the sugar.

DNA (deoxyribonucleic acid): a gigantic molecule that has the shape of a spiral staircase. DNA contains the instructions to build proteins that perform nearly all the tasks in the cell. Thus, the DNA contains the 'letter code' (constructed from the bases, guanine (G), cytosine (C), adenine (A) or thymine (T)), which codes for the instructions to build tens of thousands of protein types.

Dopamine: a neurotransmitter and also a small molecule that enables the communication between nerve cells responsible for feelings such as addiction and reward. See also *neurotransmitter*.

Drugs: chemical substances that ensure more neurotransmitters are secreted between nerve cells, as a result of which they communicate better with each other. When this happens between nerve cells responsible for pleasant feelings, then someone will feel extremely happy, and when this happens in the areas of the brain responsible for sight or respiratory regulation, then hallucinations or, with an overdose, respiratory arrest can occur.

Electron: extremely small negatively charged particle. Electrons, together with the atomic nucleus, form an atom. Electrons are 'the glue' from which atoms connect with one another in order to form molecules. See also *atom*.

Element: see *atom*.

Endorphin: a substance that plays a role in pleasant feelings. When the body creates endorphins, people feel very good or are no longer aware of pain. There are also synthetic endorphins with an equivalent effect, such as heroin or morphine.

Enzyme: a protein that accelerates a chemical reaction, such as the conversion of alcohol into a different substance or the splitting of fats, carbohydrates and proteins into their components.

Evolution: the changing of organisms' characteristics so that they are better adapted to their environment. This change can occur through random mutations in DNA, which very occasionally can have positive consequences so that an organism can survive better and reproduce better. The DNA that mutates determines which proteins are made in each cell. In turn, proteins determine the function, form and cooperation between cells that build up an entire organism.

Fat: a molecule constructed out of a 'head' and various 'tails'. These tails are long chains of carbon atoms to which hydrogen atoms are bonded. The tails are called fatty acids. When the fatty acid has one or more double bonds between two carbon atoms (a double bond means a stronger bond), then that is an unsaturated fatty acid. When the fatty acid does not have any double bonds, then that is a saturated fatty acid (in that the carbon chain is completely 'saturated' with hydrogen atoms). Saturated fats clump together easily in the body and are therefore, in some cases, unhealthier. Trans fats are constructed out of unsaturated fatty acids, of which the two hydrogen atoms are located on the opposite side of the double bond. Because of this 'strange' configuration, trans fats can be difficult to break down in the body. Trans fats are contained in industrially prepared foods such as cookies, pastry or fast food.

Fatty acid: see *fat*.

Flavonoids: these are substances that often give flowers, vegetables and fruit their red, blue or other specific colour. Flavonoids are often especially healthy for the body, usually because they are slightly toxic or because they influence specific proteins in the cell and not so much because they are 'antioxidants'.

Free radical: a free radical is a very small molecule that is very reactive. That means that a free radical very quickly enters into a chemical reaction with stable molecules in the environment, such as proteins or DNA. Thus, free radicals react with proteins, DNA or molecules in the cell walls, as a result of which they become damaged. Free radicals arise as a by-product of cell metabolism. See also *oxidation* and *antioxidant*.

Gene: a piece of the DNA strand that encodes for the creation of a specific protein. Human DNA has approximately 24,000 genes.

Glucose: glucose consists of a hexagon of carbon, oxygen and hydrogen atoms. See also *carbohydrate*.

Glycaemic index: a measure for the 'sugar peak' that a food causes in the blood. Products that are constructed of many loose or heated sugars (such as white bread or pizza dough) are broken down rapidly in the gut, as a result of which the sugars from which these products are constructed quickly end up in the blood and in this way cause high sugar peaks.

Growth hormone: see *IGF*.

Heart attack: the sudden complete clogging of a blood vessel that was already partially constricted through years of slow clogging, which occurred because cholesterol and inflammation cells accumulated in the blood vessel wall.

Homeostasis: is the process that ensures that each of the hundred thousand billion cells from which the body consists bathes in an optimal environment. Homeostasis consists of the maintaining of a constant body temperature, blood pressure, quantity of oxygen and sugar in the blood, etc. As a result, the body is constantly in contact with the environment: people breathe, eat and drink in order to keep all of those body parameters constant.

Homo neanderthalensis: see *Neanderthal*.

Homo sapiens: the human species. Arose approximately 180,000 years ago.

Hormone: a hormone can be a piece of protein, or a fatty molecule such as cholesterol or testosterone. Hormones are made in glands, such as the thyroid gland or the adrenal glands and are emitted into the bloodstream. In this way, they travel to and enter their target cells in the body where they influence the operation of the cellular machinery: more or fewer proteins are made so that the cell either gets new functions or functions are shut down.

IGF: insulin-like growth factor. IGF consists of a short chain of amino acids. IGF stimulates the cells to grow. An excess of IGF increases the chance of cancer and diabetes. Growth hormone releases IGF.

Immune system: consists of some billions of white blood cells that circulate in the tissues and blood stream and remove bacteria and viruses in the body.

Insulin: see *diabetes*.

Ion: a charged atom because it has one fewer electron, or precisely one additional electron. In all of our body fluids, ions float around and react with the charged parts of proteins in order to influence their operation. If calcium ions flow into a muscle cell, then they will adhere to long interlocking proteins, through which they become shorter. When this happens in millions of muscle cells, then muscles can become shorter in their entirety so that you can immediately turn the following page.

Macular degeneration: the dying away of the retina cells in the eye because glycated or oxidised (which means damaged) proteins accumulate in these cells. See also *protein* and *oxidation*.

Micrometre: one-billionth of one metre (10^{-6} metre) or one-thousandth of one millimetre.

Mitochondrion: a cell usually contains a couple of hundred

mitochondria, which are the energy power plants of the cell. They absorb oxygen, fats and sugars in order to produce energy-rich molecules, specifically ATP. These ATP molecules adhere to proteins, as a result of which they change structure and can therefore perform specific functions. See also *adenosine triphosphate (ATP)*.

Molecule: a molecule is when two or more atoms bond to each other. A water molecule consists of two hydrogen atoms (H) and one oxygen atom (O), while a DNA molecule consists of many millions of atoms. See also *atom*.

Mutation: a change in the DNA of a cell. The DNA contains the instructions for building proteins, implying that a mutation in the DNA results in a change in proteins, so that cells or entire bodies get different characteristics. Mutations can be caused by errors during the multiplication of cells, by (solar) radiation or by chemical substances. Mutations in a normal body cell sometimes make it possible for a cell to continue dividing itself unchecked: that is cancer. Cancer cells continue to divide themselves and ultimately proliferate throughout the entire body. See also *protein*.

Nano: a prefix that is used in order to indicate things that occur at the scale of atoms or molecules. A nano-machine is a few nanometres to a few hundreds of nanometres long. One nanometre amounts to one-millionth of a millimetre.

Nerve cell: see *neuron*.

Neuron: a brain cell or nerve cell. Neurons can send nerve signals (impulses) along their cell membrane. This occurs in the form of ions that flow into the cell through opened protein channels and in that way open up nearby channels that let in even more ions, etc.

Neurotransmitter: neurotransmitters allow the communication between nerve cells. These are molecules that are sprayed between two neurons, and in that way, stimulate the next neuron. Examples of neurotransmitters are serotonin and dopamine.

Nutrient: a healthy nutritious substance, such as a vitamin, mineral or flavonoid. See also *flavonoids*.

Oxidation: the process whereby an atom loses electrons. This can occur because a free radical 'steals' an electron from the atom in question. As a result, the atom becomes damaged. See also *antioxidant* and *free radical*.

Photon: a light particle or a quantity of light.

Placebo: a fake medicine. When the efficacy of an agent must be tested in a study, then one group gets the agent to be tested and the other group gets a placebo. This is necessary because people will already feel better when they are just given a pill, without it containing an active ingredient (this is called the 'placebo effect').

Protein: an enormous molecule that consists of anything from tens to thousands of amino acids. That way, proteins are made up of anything from hundreds to many tens of thousands of atoms. Proteins can take on all kinds of shapes and functions, and fulfil the most diverse tasks in our body. Implanted in cell membranes they function as channels, in muscle cells they can contract in order to set our muscles in motion, and in our bloodstream they transport oxygen or attack bacteria. See also *amino acid*.

Proton: the nucleus of an atom consists of neutrons and protons. Protons are relatively heavy and positively charged particles.

Quadrillion: thousand trillion (a 1 with 15 zeroes).

Saturated fat: see *fat*.

Serotonin: a neurotransmitter; and also a small molecule that provides for the communication between billions of nerve cells in the brain. See also *neurotransmitter*.

Stroke: a stroke occurs because a blood vessel in the brain tears or clogs completely, as a result of which the brain no longer gets any more blood. Because of this, the brain cells die.

Sugar peak: see *glycaemic index*.

Sugar: see *carbohydrate*.

Trans fat: see *fat*.

Unsaturated fat: see *fat*.

UV radiation: a form of energy that cannot be detected by the eye, as opposed to normal light. The radiation that the sun emits consists of normal light and UV light. The UV light contains a lot of energy and can damage the DNA in our skin cells.

White blood cells: cells that are part of the immune system and that combat bacteria and viruses that have entered the body.

References

1. Ayyadevara, S., Tazearslan, C., Bharill, P. *et al.* Caenorhabditis elegans PI3K mutants reveal novel genes underlying exceptional stress resistance and lifespan. *Aging Cell*, 2009, 8: 706–725.
2. Gems, D., Sutton, A.J., Sundermeyer, M.L. *et al.* Two pleiotropic classes of daf-2 mutation affect larval arrest, adult behavior, reproduction and longevity in Caenorhabditis elegans. *Genetics*, 1998, 150: 129–155.
3. Wyndaele, J.J. *Urinewegen* (p 230). 2008, Universiteit Antwerpen.
4. Seddon, J.M., Ajani, U.A., Sperduto, R.D. *et al.* Dietary carotenoids, vitamins A, C, and E, and advanced age-related macular degeneration. Eye Disease Case-Control Study Group. JAMA, 1994, 272: 1413–1420.
5. Seddon, J.M., George, S., en Rosner, B. Cigarette smoking, fish consumption, omega-3 fatty acid intake, and associations with age-related macular degeneration: the US Twin Study of Age Related Macular Degeneration. *Arch. Ophthalmol.*, 2006, 1241 995–1001.
6. Bliznakov, E.G. *Biomedical and clinical aspects of co-enzym Q10. 3* (p 311). 1981, Elsevier, Amsterdam.
7. de Grey, Aubrey en Rae, Michael. *Ending Aging.* p 173. 2007. St. Martin's Griffin.
8. Moyer, Melinda Werner. Carbs against cardio. 1–5–2010, *Scientific American*.
9. Austad, S.N. Methusaleh's Zoo: how nature provides us with clues for extending human health span. *J. Comp Pathol.*, 2010, 142 Suppl 1: S10–S21.
10. Gecommentarieerd Geneesmiddelen Repertorium. 2011. BCFI.

11. Cosgrove, M.C., Franco, O.H., Granger, S.P., Murray, P.G., and Mayes, A.E. Dietary nutrient intakes and skin-aging appearance among middle-aged American women. *Am. J. Clin. Nutr.*, 2007, 86: 1225–1231.

12. Hankinson, S.E., Willett, W.C., Colditz, G.A. *et al.* Circulating concentrations of insulin-like growth factor 1 and risk of breast cancer. *Lancet*, 1998, 351: 1393–1396.

13. Stattin, P., Bylund, A., Rinaldi, S. *et al.* Plasma insulin-like growth factor 1, insulin-like growth factor-binding proteins, and prostate cancer risk: a prospective study. *J. Natl. Cancer Inst.*, 2000, 92: 1910–1917.

14. Chan, J.M., Stampfer, M.J., Ma, J. *et al.* Insulin-like growth factor 1 (IGF-1) and IGF binding protein-3 as predictors of advanced-stage prostate cancer. *J. Natl. Cancer Inst.*, 2002, 94: 1099–1106.

15. Santisteban, G.A., Ely, J.T., Hamel, E.E., Read, D.H., and Kozawa, S.M. Glycemic modulation of tumor tolerance in a mouse model of breast cancer. *Biochem. Biophys. Res. Commun.*, 1985, 132: 1174–1179.

16. Guevara-Aguirre, J., Balasubramanian, P., Guevara-Aguirre, M. *et al.* Growth hormone receptor deficiency is associated with a major reduction in pro-aging signaling, cancer, and diabetes in humans. *Sci. Transl. Med.*, 2011, 3: 70ra13.

17. Green, J., Cairns, B.J., Casabonne, D. *et al.* Height and cancer incidence in the Million Women Study: prospective cohort, and meta-analysis of prospective studies of height and total cancer risk. *Lancet Oncol.*, 2011, 12: 785–794.

18. Pelicano, H., Martin, D.S., Xu, R.H., and Huang, P. Glycolysis inhibition for anticancer treatment. *Oncogene*, 2006, 25: 4633–4646.

19. Hotchkiss, R. and Karl, I.E. The pathophsyiology and treatment of sepsis. *New England Journal of Medicine*, 2003, 348: 138–150.

20. Lim, E.L., Hollingsworth, K.G., Aribisala, B.S. *et al.* Reversal of type 2 diabetes: normalisation of beta cell function in association with decreased pancreas and liver triacylglycerol. *Diabetologia*, 2011, 54: 2506–2514.

21. Diabetes Prevention Program Research Group, Knowler W.C.,

Fowler S.E., Hamman R.F., Christophi C.A. *et al.* 10-year follow-up of diabetes incidence and weight loss in the Diabetes Prevention Program Outcomes Study. *Lancet*, 2009 Nov. 14; 374(9702): 1677–86.

22. Lammert, A., Kratzsch J., Selhorst J., Humpert P.M. *et al.* Clinical benefit of a short term dietary oatmeal intervention in patients with type 2 diabetes and severe insulin resistance: a pilot study. *Exp Clin Endocrinol Diabetes*, 2008 Feb; 116 (2): 132–4.

23. Hyman, M., *The blood sugar solution*, Little, Brown and Company, 2012.

24. Intensive blood-glucose control with sulphonylureas or insulin compared with conventional treatment and risk of complications in patients with type-2-diabetes (ukpds 33). UK Prospective Diabetes Study (UKPDS) *Group. Lancet*, 1998, 352: 837–853.

25. Ericson, U., Sonestedt E., Gullberg B., Hellstrand S. *et al.* High intakes of protein and processed meat associate with increased incidence of type 2 diabetes. *Br. J. Nutr.*, 2012 Aug. 1:1–11.

26. Grandison, R.C., Piper, M.D., and Partridge, L. Amino-acid imbalance explains extension of lifespan by dietary restriction in Drosophila. *Nature*, 2009, 462: 1061–1064.

27. Ross, M.H. and Bras, G. Dietary preference and diseases of age. *Nature*, 1974, 250: 263–265.

28. Walford, Roy. *Beyond the 120 year diet.* 2000, Thunder's Mouth Press, New York.

29. Ericson, U., Sonestedt E., Gullberg B., Hellstrand S. *et al.* High intakes of protein and processed meat associate with increased incidence of type 2 diabetes. *Br. J. Nutr.*, 2012 Aug. 1:1–11.

30. Walford, Roy. *Beyond the 120 year diet.* 2000, Thunder's Mouth Press, New York.

31. Cho, E., Chen, W.Y., Hunter, D.J. *et al.* Red meat intake and risk of breast cancer among premenopausal women. *Arch. Intern. Med.*, 2006, 166: 2253–2259.

32. Willett, W.C., Stampfer, M.J., Colditz, G.A., Rosner, B.A., and Speizer, F.E. Relation of meat, fat, and fiber intake to the risk of colon cancer in a prospective study among women. N. *Engl. J. Med.*, 1990, 323: 1664–1672.

33. Pan, A. *et al.* Red meat consumption and mortality: results from

2 prospective cohort studies. *Archives of internal medicine,* 2012, 172(7), 555–63.

34. Rohrmann, S. et al. Meat consumption and mortality – results from the European Prospective Investigation into Cancer and Nutrition. bmc *Medicine,* 2013 11(1), 63.

35. Fasano, Alessio. Surprises from celiac disease. 1–8–2009, *Scientific American.*

36. Harrison, D.E., Strong, R., Sharp, Z.D. *et al.* Rapamycin fed late in life extends lifespan in genetically heterogeneous mice. *Nature,* 2009, 460: 392–395.

37. Song, Y., Manson, J.E., Buring, J.E., and Liu, S. A prospective study of red meat consumption and type-2-diabetes in middle-aged and elderly women: the women's health study. *Diabetes Care,* 2004, 27: 2108–2115.

38. Buettner, Dan. *Het geheim van langer leven.* 2009. National Geographic Books (in cooperation with Uitgeverij Carrera).

39. Tremblay, F. *et al.* Overactivation of S6 kinase 1 as a cause of human insulin resistance during increased amino acid availability. Diabetes, 2005, 2674–2684.

40. Krebs, M. *et al.* Amino acid-dependent modulation of glucose metabolism in humans. *Eur. J. Clin. Invest,* 2005. 35, 351–354.

41. Takahiro Nobukuni *et al.* Amino acids mediate mtor/raptor signaling through activation of class 3 phosphatidylinositol 3OH-kinase. *Proc Natl Acad Sci.,* 2005 October 4; 102(40): 14238–14243.

42. Li S., Ogawa W., Emi A., Hayashi K. *et al.* Role of S6K1 in regulation of srebp1c expression in the liver. *Biochem Biophys Res Commun.,* 2011 Aug 26; 412(2): 197–202.

43. Siri-Tarino, P.W., Sun, Q., Hu, F.B., and Krauss, R.M. Meta-analysis of prospective cohort studies evaluating the association of saturated fat with cardiovascular disease. *Am. J. Clin. Nutr.,* 2010, 91: 535–546.

44. Iris, S., Sccwarzfuchs, D., Henkin, Y., *et al.* Weight Loss with a Low-Carbohydrate, Mediterranean, or Low-Fat Diet. 2008, 359: 229–241.

45. Thomas DE, Elliott EJ, Baur L. Low glycaemic index or low glycaemic load diets for overweight and obesity. *Cochrane Database Syst Rev,* 2007.

46. Ebbeling C.B. *et al.* Effects of dietary composition on energy expenditure during weight-loss maintenance. *JAMA*, 2012 Jun 27; 307(24): 2627–34.

47. Beulens, J.W., de Bruijne, L.M., Stolk, R.P. *et al.* High dietary glycemic load and glycemic index increase risk of cardiovascular disease among middle-aged women: a population-based follow-up study. *J. Am. Coll. Cardiol.*, 2007, 50: 14–21.

48. Weingartner, O., Bohm, M., and Laufs, U. Controversial role of plant sterol esters in the management of hypercholesterolaemia. *Eur. Heart J.*, 2009, 30: 404–409.

49. Wang, C., Harris, W.S., Chung, M. *et al.* n-3 Fatty acids from fish or fish-oil supplements, but not alpha-linolenic acid, benefit cardiovascular disease outcomes in primary- and secondary-prevention studies: a systematic review. *Am. J. Clin. Nutr.*, 2006, 84: 5–17.

50. Lee, J.H., O'Keefe, J.H. en et al. Omega-3 Fatty Acids for Cardioprotection. 83 (3), 324–332. 2008, Mayo Clinic Proceedings.

51. Marchioli, R., Barzi, F., Bomba, E. *et al.* Early protection against sudden death by n-3 polyunsaturated fatty acids after myocardial infarction: time-course analysis of the results of the Gruppo Italiano per lo Studio della Sopravvivenza nell'Infarto Miocardico (gissi)-Prevenzi-one. *Circulation*, 2002, 105: 1897–1903.

52. Kris-Etherton, P.M., Harris, W.S. and Appel, L.J. Fish consumption, fish oil, omega-3 fatty acids, and cardiovascular disease. *Circulation*, 2002, 106: 2747–2757.

53. De, B.G., Ambrosioni, E., Borch-Johnsen, K. *et al.* European guide-lines on cardiovascular disease prevention in clinical practice. Third Joint Task Force of European and other Societies on Cardiovascular Disease Prevention in Clinical Practice (constituted by representatives of eight societies and by invited experts). *Atherosclerosis*, 2004, 173: 381–391.

54. Albert, C.M., Campos, H., Stampfer, M.J. *et al.* Blood levels of long-chain n-3 fatty acids and the risk of sudden death. *N. Engl. J. Med.*, 2002, 346: 1113–1118.

55. Sinzinger, H. and O'Grady, J. Professional athletes suffering from familial hypercholesterolaemia rarely tolerate statin

treatment because of muscular problems. *Br. J. Clin. Pharmacol.*, 2004, 57: 525–528.

56. Gaist, D., Jeppesen, U., Andersen, M. *et al.* Statins and risk of polyneuropathy: a case-control study. *Neurology*, 2002, 58: 1333–1337.

57. Studer, M., Briel, M., Leimenstoll, B., Glass, T.R., and Bucher, H.C. Effect of different antilipidemic agents and diets on mortality: a systematic review. *Arch. Intern. Med.*, 2005, 165: 725–730.

58. Harris, W.S. Omega-3 fatty acids and cardiovascular disease: a case for omega-3 index as a new risk factor. *Pharmacol. Res.*, 2007, 55: 217–223.

59. Din, J.N., Harding, S.A., Valerio, C.J. *et al.* Dietary intervention with oil rich fish reduces platelet-monocyte aggregation in man. *Atherosclerosis*, 2008, 197: 290–296.

60. Reiffel, J.A. and McDonald, A. Antiarrhythmic effects of omega-3 fatty acids. *Am. J. Cardiol.*, 2006, 98: 50i-60i.

61. Calo, L., Bianconi, L., Colivicchi, F. *et al.* n-3 Fatty acids for the prevention of atrial fibrillation after coronary artery bypass surgery: a randomized, controlled trial. *J. Am. Coll. Cardiol.*, 2005, 45: 1723–1728.

62. Stoll, A.L., Severus, W.E., Freeman, M.P. *et al.* Omega-3 fatty acids in bipolar disorder: a preliminary double-blind, placebo-controlled trial. *Arch. Gen. Psychiatry*, 1999, 56: 407–412.

63. Nemets, B., Stahl, Z., and Belmaker, R.H. Addition of omega-3 fatty acid to maintenance medication treatment for recurrent unipolar depressive disorder. *Am. J. Psychiatry*, 2002, 159: 477–479.

64. Lesperance, F., Frasure-Smith, N., St-Andre, E. *et al.* The efficacy of omega-3 supplementation for major depression: a randomized controlled trial. *J. Clin. Psychiatry*, 2011, 72: 1054–1062.

65. Amminger, G.P., Schafer, M.R., Papageorgiou, K. *et al.* Long-chain omega-3 fatty acids for indicated prevention of psychotic disorders: a randomized, placebo-controlled trial. *Arch. Gen. Psychiatry*, 2010, 67: 146–154.

66. Al, M.D., van Houwelingen, A.C., and Hornstra, G. Long-chain poly-unsaturated fatty acids, pregnancy, and pregnancy outcome. *Am. J. Clin. Nutr.*, 2000, 71: 285S-291S.

67. Hibbeln, J.R., Davis, J.M., Steer, C. *et al.* Maternal seafood consumption in pregnancy and neurodevelopmental outcomes in childhood (alspac study): an observational cohort study. *Lancet*, 2007, 369: 578–585.

68. Schaefer, E.J., Bongard, V., Beiser, A.S. *et al.* Plasma phosphatidylcholine docosahexaenoic acid content and risk of dementia and Alzheimer disease: the Framingham Heart Study. *Arch. Neurol.*, 2006, 63: 1545–1550.

69. Schatzberg, A. et al. *Manual of clinical psychopharmacology* (358). 2011, American Psychiatric Publishing.

70. Horrobin, D.F. Lipid metabolism, human evolution and schizophrenia. Prostaglandins Leukot. *Essent. Fatty Acids*, 1999, 60: 431–437.

71. Marean, Curtis W. When the Sea Saved Humanity. *Scientific American*, 2010.

72. Volker, D., Fitzgerald, P., Major, G. and Garg, M. Efficacy of fish oil concentrate in the treatment of rheumatoid arthritis. *J. Rheumatol.*, 2000, 27: 2343–2346.

73. Lau, C.S., Morley, K.D. and Belch, J.J. Effects of fish oil supplementation on non-steroidal anti-inflammatory drug requirement in patients with mild rheumatoid arthritis – a double-blind placebo controlled study. *Br. J. Rheumatol.*, 1993, 32: 982–989.

74. Broughton, K.S., Johnson, C.S., Pace, B.K., Liebman, M. and Kleppinger, K.M. Reduced asthma symptoms with n 3 fatty acid ingestion are related to 5-series leukotriene production. *Am. J. Clin. Nutr.*, 1997, 65: 1011–1017.

75. Belluzzi, A., Brignola, C., Campieri, M. *et al.* Effect of an enteric-coated fish-oil preparation on relapses in Crohn's disease. *N. Engl. J. Med.*, 1996, 334: 1557–1560.

76. USDA Agricultural Research Service. Nutrient Data Laboratory. www.ars.usda.gov/nutrientdata. 2011.

77. Spiteller, G. Furan fatty acids: occurrence, synthesis, and reactions. Are furan fatty acids responsible for the cardioprotective effects of a fish diet? *Lipids*, 2005, 40: 755–771.

78. Foran, J.A., Hites, R.A., Carpenter, D.O. *et al.* A survey of metals in tissues of farmed Atlantic and wild Pacific salmon. *Environ. Toxicol. Chem.*, 2004, 23: 2108–2110.

79. Foran, S.E., Flood, J.G. and Lewandrowski, K.B. Measurement of mercury levels in concentrated over-the-counter fish oil preparations: is fish oil healthier than fish? *Arch. Pathol. Lab Med.*, 2003, 127: 1603–1605.

80. Jacqueline Chan et al. Water, Other Fluids, and Fatal Coronary Heart Disease – The Adventist Health Study. *American Journal of Epidemiology*, 2002.

81. Hattori M, Azami Y. Searching for preventive measures of cardiovascular events in aged Japanese taxi drivers, *J Hum Ergol (Tokyo)*, 2001 Dec; 30(1–2): 321–6.

82. Thring, T.S., Hili, P. and Naughton, D.P. Anti-collagenase, anti-elastase and anti-oxidant activities of extracts from 21 plants. bmc. *Complement Altern. Med.*, 2009, 9: 27.

83. Bjelakovic, G., Nikolova, D., Gluud, L.L., Simonetti, R.G. and Gluud, C. Mortality in randomized trials of antioxidant supplements for primary and secondary prevention: systematic review and meta-analysis. JAMA, 2007, 297: 842–857.

84. Khafif, A., Schantz, S.P., al-Rawi, M., Edelstein, D. and Sacks, P.G. Green tea regulates cell cycle progression in oral leukoplakia. *Head Neck*, 1998, 20: 528–534.

85. Hastak, K., Agarwal, M.K., Mukhtar, H. and Agarwal, M.L. Ablation of either p21 or Bax prevents p53-dependent apoptosis induced by green tea polyphenol epigallocatechin-3-gallate. FASEB j., 2005, 19: 789–791.

86. Fujiki, H., Suganuma, M., Kurusu, M. *et al.* New TNF-alpha releasing inhibitors as cancer preventive agents from traditional herbal medicine and combination cancer prevention study with egcg and sulindac or tamoxifen. *Mutat. Res.*, 2003, 523–524: 119–125.

87. De, F.S. Mechanisms of inhibitors of mutagenesis and carcinogenesis. *Mutat. Res.*, 1998, 402: 151–158.

88. Imai, K., Suga, K., and Nakachi, K. Cancer-preventive effects of drinking green tea among a Japanese population. *Prev. Med.*, 1997, 26: 769–775.

89. Yang, G., Shu, X.O., Li, H. *et al.* Prospective cohort study of green tea consumption and colorectal cancer risk in women. Cancer Epidemiol. *Biomarkers Prev.*, 2007, 16: 1219–1223.

90. Kurahashi, N., Sasazuki, S., Iwasaki, M., Inoue, M. and Tsugane,

S. Green tea consumption and prostate cancer risk in Japanese men: a prospective study. *Am. J. Epidemiol.*, 2008, 167: 71–77.

91. Meltzer, S.M., Monk, B.J. and Tewari, K.S. Green tea catechins for treatment of external genital warts. *Am. J. Obstet. Gynecol.*, 2009, 200: 233–237.

92. Kao, Y.H., Chang, H.H., Lee, M.J. and Chen, C.L. Tea, obesity, and diabetes. *Mol. Nutr. Food Res.*, 2006, 50: 188–210.

93. Venables, M.C., Hulston, C.J., Cox, H.R. and Jeukendrup, A.E. Green tea extract ingestion, fat oxidation, and glucose tolerance in healthy humans. *Am. J. Clin. Nutr.*, 2008, 87: 778–784.

94. Arab, L., Liu, W., and Elashoff, D. Green and black tea consumption and risk of stroke: a meta-analysis. *Stroke*, 2009, 40: 1786–1792.

95. Funk, J.L., Frye, J.B., Oyarzo, J.N. and Timmermann, B.N. Comparative effects of two gingerol-containing Zingiber officinale extracts on experimental reumatoid artritis. *J. Nat. Prod.*, 2009, 72: 403–407.

96. Zick, S.M., Turgeon, D.K., Vareed, S.K. *et al*. Phase II Study of the Effects of Ginger Root Extract on Eicosanoids in Colon Mucosa in People at Normal Risk for Colorectal Cancer. *Cancer Prev. Res. (Phila)*, 2011, 4: 1929–1937.

97. Lee, H.S., Seo, E.Y., Kang, N.E. and Kim, W.K. [6]-Gingerol inhibits metastasis of MDA-MB-231 human breast cancer cells. *J. Nutr. Biochem.*, 2008, 19: 313–319.

98. Willett, W.C. and Skerrett, P.J. *Eat, drink and be healthy: The Harvard Medical School Guide to Healthy Eating*. 2005, Free Press.

99. Ohnishi, M. and Razzaque, M.S. Dietary and genetic evidence for phosphate toxicity accelerating mammalian aging. FASEB j., 2010, 24: 3562–3571.

100. Meeting of the American Diabetes Association, San Diego, California, June 26, 2011.

101. Nettleton, J.A., Lutsey P.L., Wang Y., Lima J.A *et al*. Diet soda intake and risk of incident metabolic syndrome and type 2 diabetes in the Multi-Ethnic Study of Atherosclerosis (mesa). *Diabetes Care*, 2009 Apr; 32(4): 688–94.

102. Popkin, B.M. The world is fat. *Sci. Am.*, 2007, 297: 88–95.

103. Global Market Information Database, Euromonitor. 2011.
104. Nakamura, K., Iwahashi, K., Furukawa, A. *et al.* Acetaldehyde adducts in the brain of alcoholics. *Arch. Toxicol.*, 2003, 77: 591–593.
105. Pearson, K.J., Baur, J.A., Lewis, K.N. *et al.* Resveratrol delays age-related deterioration and mimics transcriptional aspects of dietary restriction without extending life span. *Cell Metab*, 2008, 8: 157–168.
106. Baur, J.A., Pearson, K.J., Price, N.L. *et al.* Resveratrol improves health and survival of mice on a high-calorie diet. *Nature*, 2006, 444: 337–342.
107. Eskelinen, M.H., Ngandu, T., Tuomilehto, J., Soininen, H. and Kivipelto, M. Midlife coffee and tea drinking and the risk of late-life dementia: a population-based CAIDE study. *J. Alzheimers. Dis.*, 2009, 16: 85–91.
108. Ross, G.W., Abbott, R.D., Petrovitch, H. *et al.* Association of coffee and caffeine intake with the risk of Parkinson disease. JAMA, 2000, 283: 2674–2679.
109. Salazar-Martinez, E., Willett, W.C., Ascherio, A. *et al.* Coffee consumption and risk for type-2-diabetes mellitus. *Ann. Intern. Med.*, 2004, 140: 1–8.
110. Butt, M.S. and Sultan, M.T. Coffee and its consumption: benefits and risks. *Crit Rev. Food Sci. Nutr.*, 2011, 51: 363–373.
111. Verhoef, P., Pasman, W.J., Van, V.T., Urgert, R. and Katan, M.B. Contribution of caffeine to the homocysteine-raising effect of coffee: a randomized controlled trial in humans. *Am. J. Clin. Nutr.*, 2002, 76: 1244–1248.
112. Park, M., Ross, G.W., Petrovitch, H. *et al.* Consumption of milk and calcium in midlife and the future risk of Parkinson disease. *Neurology*, 2005, 64: 1047–1051.
113. Fairfield, K. Annual Meeting of the Society for General Internal Medicine: Diary products linked to ovarian cancer risk. *Family Practice News*, 2000.
114. Chan, J.M., Stampfer, M.J., Ma, J. *et al.* Dairy products, calcium, and prostate cancer risk in the Physicians' Health Study. *Am. J. Clin. Nutr.*, 2001, 74: 549–554.
115. Qin LQ, He K, Xu JY.. Milk consumption and circulating

insulin-like growth factor-I level: a systematic literature review. Int J Food Sci Nutr. 2009.

116. Maggi, S., Kelsey, J.L., Litvak, J. and Heyse, S.P. Incidence of hip fractures in the elderly: a cross-national analysis. *Osteoporos. Int.*, 1991, 1: 232–241.

117. Hegsted, D.M. Calcium and osteoporosis. *J. Nutr.*, 1986, 116: 2316–2319.

118. Feskanich, D., Willett, W.C., Stampfer, M.J. and Colditz, G.A. Milk, dietary calcium, and bone fractures in women: a 12-year prospective study. *Am. J. Public Health*, 1997, 87: 992–997.

119. Tucker, K.L., Hannan, M.T., Chen, H. *et al.* Potassium, magnesium, and fruit and vegetable intakes are associated with greater bone mineral density in elderly men and women. *Am. J. Clin. Nutr.*, 1999, 69: 727–736.

120. Chandalia, M., Garg A, Lutjohann D., von Bergmann K., Grundy S.M., Brinkley L.J. *et al.* Beneficial effects of high dietary fiber intake in patients with type 2 diabetes mellitus. *N. Engl. J. Med.*, 2000 May 11;342(19):1392–8.

121. Lammert, A., Kratzsch J., Selhorst J., Humpert P.M. *et al.* Clinical benefit of a short term dietary oatmeal intervention in patients with type 2 diabetes and severe insulin resistance: a pilot study. *Exp. Clin. Endocrinol. Diabetes.*, 2008 Feb; 116(2): 132–4.

122. Bliss, R.M., Oats: cooling inflammation and unhealthy cell proliferation. *Agricultural Research*, 2010.

123. Meydani, M. Potential health benefits of avenanthramides of oats. *Nutr. Rev.*, 2009, 67(12): 731–5.

124. European Food Safety Authority. Scientific Opinion on the substantiation of health claims related to beta-glucans and maintenance of normal blood cholesterol concentrations and maintenance or achievement of a normal body weight. *EFSA Journal*, 2009, 7(9):1254.

125. Cosgrove, M.C., Franco, O.H., Granger, S.P., Murray, P.G. and Mayes, A.E. Dietary nutrient intakes and skin-aging appearance among middle-aged American women. *Am. J. Clin. Nutr.*, 2007, 86: 1225–1231.

126. Heaton, K.W., Marcus S.N., Emmett P.M., Bolton C.H. Particle size of wheat, maize, and oat test meals: effects on

plasma glucose and insulin responses and on the rate of starch digestion in vitro. *Am. J. Clin. Nutr.*, 1988, 47(4): 675–82

127. Rasmussen, O., Winther, E., Heransen, K. Postprandial glucose and insulin responses to rolled oats ingested raw, cooked or as a mixture with raisins in normal subjects and type 2 diabetic patients. *Diabetic Medicine*, 1989.

128. Miller, R.A., Buehner, G., Chang, Y. *et al*. Methionine-deficient diet extends mouse lifespan, slows immune and lens aging, alters glucose, T4, IGF-1 and insulin levels, and increases hepatocyte MIF levels and stress resistance. *Aging Cell*, 2005, 4: 119–125.

129. Qin, L.Q., Xu, J.Y., Wang, P.Y. and Hoshi, K. Soyfood intake in the prevention of breast cancer risk in women: a meta-analysis of observational epidemiological studies. *J. Nutr. Sci. Vitaminol.* (Tokyo), 2006, 52: 428–436.

130. Trock, B.J., Hilakivi-Clarke, L. and Clarke, R. Meta-analysis of soy intake and breast cancer risk. *J. Natl. Cancer Inst.*, 2006, 98: 459–471.

131. Hamilton-Reeves, J.M., Vazquez G., Duval S.J., *et al*. Clinical studies show no effects of soy protein or isoflavones on reproductive hormones in men: results of a meta-analysis. *Fertil. Steril.* 2010;94:997–1007.

132. Messina, M., Watanabe, S. and Setchell, K.D. Report on the 8th International Symposium on the Role of Soy in Health Promotion and Chronic Disease Prevention and Treatment. *J. Nutr.*, 2009, 139: 796S-802S.

133. White, L.R., Petrovitch, H., Ross, G.W. *et al*. Brain aging and midlife tofu consumption. *J. Am. Coll. Nutr.*, 2000, 19: 242–255.

134. Hogervorst, E., Sadjimim, T., Yesufu, A., Kreager, P. en Rahardjo, T.B. High tofu intake is associated with worse memory in elderly Indonesian men and women. *Dement. Geriatr. Cogn Disord.*, 2008, 26: 50–57.

135. Kim, J.Y., Gum S.N., Paik J.K., Lim H.H. *et al*. Effects of nattokinase on blood pressure: a randomized, controlled trial. *Hypertens. Res.*, 2008 Aug;31(8):1583–8.

136. Fujita, M., Nomura K., Hong K., Ito Y., Asada A., Nishimuro S. Purification and characterization of a strong fibrinolytic

enzyme (nattokinase) in the vegetable cheese natto, a popular soybean fermented food in Japan. *Biochem. Biophys. Res. Commun.*, 1993; 197: 1340–1347.

137. Hsu, R.L., Lee K.T., Wang J.H., Lee L.Y. *et al.* Amyloid-degrading ability of nattokinase from Bacillus subtilis natto. *J. Agric. Food. Chem.*, 2009 Jan 28;57(2):503–8.

138. Torisu, M., Hayashi, Y., Ishimitsu, T. *et al.* Significant prolongation of disease-free period gained by oral polysaccharide K (PSK) administration after curative surgical operation of colorectal cancer. *Cancer Immunol. Immunother.*, 1990, 31: 261–268.

139. Nakazato, H., Koike, A., Saji, S., Ogawa, N., and Sakamoto, J. Efficacy of immunochemotherapy as adjuvant treatment after curative resection of gastric cancer. Study Group of Immuno-chemotherapy with psk for Gastric Cancer. *Lancet*, 1994, 343: 1122–1126.

140. Hara, M., Hanaoka, T., Kobayashi, M. *et al.* Cruciferous vegetables, mushrooms, and gastrointestinal cancer risks in a multicenter, hospital-based case-control study in Japan. *Nutr. Cancer*, 2003, 46: 138–147.

141. Zhang, M., Huang, J., Xie, X., and Holman, C. D. Dietary intakes of mushrooms and green tea combine to reduce the risk of breast cancer in Chinese women. *Int. J. Cancer*, 2009, 124: 1404–1408.

142. Béliveau, R. and D. Gingras. *Eten tegen kanker* 2011, Kosmos Uitgevers.

143. Goodman, M.T., Kiviat, N., McDuffie, K. *et al.* The association of plasma micronutrients with the risk of cervical dysplasia in Hawaii. *Cancer Epidemiol. Biomarkers Prev.*, 1998, 7: 537–544.

144. Cohen, J.H., Kristal, A.R. and Stanford, J.L. Fruit and vegetable intakes and prostate cancer risk. *J. Natl. Cancer Inst.*, 2000, 92: 61–68.

145. Ambrosone, C.B., McCann, S.E., Freudenheim, J.L. *et al.* Breast cancer risk in premenopausal women is inversely associated with consumption of broccoli, a source of isothiocyanates, but is not modified by GST genotype. *J. Nutr.*, 2004, 134: 1134–1138.

146. Ahmed, F. Health: Edible advice. *Nature*, 2010, 468: S10–S12.
147. Bruni, F., The Billionaire Who Is Planning His 125th Birthday, *TheNew York Times*, 3-3-2011.
148. Agudo, A., Cabrera, L., Amiano, P. *et al.* Fruit and vegetable intakes, dietary antioxidant nutriënts, and total mortality in Spanish adults: findings from the Spanish cohort of the European Prospective Investigation into Cancer and Nutrition (EPIC-Spain). *Am. J. Clin. Nutr.*, 2007, 85: 1634–1642.
149. Bae, J.Y., Choi, J.S., Kang, S.W. *et al.* Dietary compound ellagic acid alleviates skin wrinkle and inflammation induced by UV-B irradiation. *Exp. Dermatol.*, 2010, 19: e182–e190.
150. Evrengul, H., Dursunoglu, D., Kaftan, A. *et al.* Bilateral diagonal earlobe crease and coronary artery disease: a significant association. *Dermatology*, 2004, 209: 271–275.
151. Aviram, M., Rosenblat, M., Gaitini, D. *et al.* Pomegranate juice consumption for 3 years by patients with carotid artery stenosis reduces common carotid intima-media thickness, blood pressure and LDL oxidation. *Clin. Nutr.*, 2004, 23: 423–433.
152. Petersen, R.C., Thomas, R.G., Grundman, M. *et al.* Vitamin E and donepezil for the treatment of mild cognitive impairment. *N. Engl. J. Med.*, 2005, 352: 2379–2388.
153. O'Byrne, D.J., Devaraj, S., Grundy, S.M., and Jialal, I. Comparison of the antioxidant effects of Concord grape juice flavonoids alpha-tocopherol on markers of oxidative stress in healthy adults. *Am. J. Clin. Nutr.*, 2002, 76: 1367–1374.
154. Duffy, K.B., Spangler, E.L., Devan, B.D. *et al.* A blueberry-enriched diet provides cellular protection against oxidative stress and reduces a kainate-induced learning impairment in rats. *Neurobiol. Aging*, 2008, 29: 1680–1689.
155. Shukitt-Hale, B., Carey, A.N., Jenkins, D., Rabin, B.M., and Joseph, J.A. Beneficial effects of fruit extracts on neuronal function and behavior in a rodent model of accelerated aging. *Neurobiol. Aging*, 2007, 28: 1187–1194.
156. Ferguson, P.J., Kurowska, E., Freeman, D.J., Chambers, A.F., and Koropatnick, D.J. A flavonoid fraction from cranberry extract inhibits proliferation of human tumor cell lines. *J. Nutr.*, 2004, 134: 1529–1535.
157. Van de Velde *et al. Oncologie.* 2005, Bohn Stafleu van Loghum.

158. Knekt, P., Kumpulainen, J., Jarvinen, R. *et al.* Flavonoid intake and risk of chronic diseases. *Am. J. Clin. Nutr.*, 2002, 76: 560–568.

159. Pan, A. *et al.* Red meat consumption and mortality: results from 2 prospective cohort studies. *Archives of internal medicine*, 2012, 172(7), 555–63.

160. Weill, P., Schmitt, B., Chesneau, G. *et al.* Effects of introducing linseed in livestock diet on blood fatty acid composition of consumers of animal products. *Ann. Nutr. Metab*, 2002, 46: 182–191.

161. Simopoulos, A.P. and Salem, N., Jr. n-3 fatty acids in eggs from range-fed Greek chickens. *N. Engl. J. Med.*, 1989, 321: 1412.

162. Our big pig problem. *Scientific American*, 1–4–2011.

163. Qureshi, A.I., Suri, F.K., Ahmed, S. *et al.* Regular egg consumption does not increase the risk of stroke and cardiovascular diseases. *Med. Sci. Monit.*, 2007, 13: cr1–cr8.

164. Gast, G.C., de Roos, N.M., Sluijs, I. *et al.* A high menaquinone intake reduces the incidence of coronary heart disease. *Nutr. Metab Cardiovasc. Dis.*, 2009, 19: 504–510.

165. Belin, R.J., Greenland, P., Martin, L. *et al.* Fish intake and the risk of incident heart failure: the Women's Health Initiative. *Circ. Heart Fail.*, 2011, 4: 404–413.

166. Grassi, D., Necozione, S., Lippi, C. *et al.* Cocoa reduces blood pressure and insulin resistance and improves endothelium dependent vasodilation in hypertensives. *Hypertension, 2005*, 46: 398–405.

167. Holt, R.R., Schramm, D.D., Keen, C.L., Lazarus, S.A., and Schmitz, H.H. Chocolate consumption and platelet function. jama, 2002, 287: 2212–2213.

168. Bayard, V., Chamorro, F., Motta, J. en Hollenberg, N.K. Does flavanol intake influence mortality from nitric oxide-dependent processes? Ischemic heart disease, stroke, diabetes mellitus, and cancer in Panama. *Int. J. Med. Sci.*, 2007, 4: 53–58.

169. Buitrago-Lopez, A., Sanderson, J., Johnson, L. *et al.* Chocolate consumption and cardiometabolic disorders: systematic review and meta-analysis. BMJ, 2011, 343: d4488.

170. Arumugam, M., Raes, J., Pelletier, E. *et al.* Enterotypes of the human gut microbiome. *Nature*, 2011, 473: 174–180.

171. Guarner, F. and Malagelada, J.R. Gut flora in health and disease. *Lancet*, 2003, 361: 512–519.

172. DeWeerdt, S. Food: The omnivore's labyrinth. *Nature*, 2011, 471: S22–S24.

173. Wierdsma, N.J., van Bodegraven, A.A., Uitdehaag, B.M. *et al.* Fructo-oligosaccharides and fibre in enteral nutrition has a beneficial influence on microbiota and gastrointestinal quality of life. *Scand. J. Gastroenterol.*, 2009, 44: 804–812.

174. Bouhnik, Y., Achour, L., Paineau, D. *et al.* Four-week short chain fructo-oligosaccharides ingestion leads to increasing fecal bifidobacteria and cholesterol excretion in healthy elderly volunteers. *Nutr. J.*, 2007, 6: 42.

175. Abha Chauhanemail et al. Walnut-rich diet improves memory deficits and learning skills in transgenic mouse model of Alzheimer's disease. *Alzheimer's & Dementia: The Journal of the Alzheimer's Association* 6 (4), 69. 1–7–2010.

176. Hu, F.B. and Stampfer, M.J. Nut consumption and risk of coronary heart disease: a review of epidemiologic evidence. *Curr. Atheroscler. Rep.*, 1999, 1: 204–209.

177. Wagner, K.H., Kamal-Eldin, A. en Elmadfa, I. Gamma-tocopherol – an underestimated vitamin? *Ann. Nutr. Metab.*, 2004, 48: 169–188.

178. Berry, E.M., Eisenberg, S., Haratz, D. *et al.* Effects of diets rich in monounsaturated fatty acids on plasma lipoproteins – the Jerusalem Nutrition Study: high mufas vs high pufas. *Am. J. Clin. Nutr.*, 1991, 53: 899–907.

179. Ros, E., Nunez, I., Perez-Heras, A. *et al.* A walnut diet improves endothelial function in hypercholesterolemic subjects: a randomized crossover trial. *Circulation*, 2004, 109: 1609–1614.

180. Dubnov, G. en Berry, E.M. Omega-6/omega-3 fatty acid ratio: the Israeli paradox. *World Rev. Nutr. Diet.*, 2003, 92: 81–91.

181. Degirolamo, C. and Rudel, L.L. Dietary monounsaturated fatty acids appear not to provide cardioprotection. *Curr. Atheroscler. Rep.*, 2010, 12: 391–396.

182. Purba, M.B., Kouris-Blazos, A., Wattanapenpaiboon, N. *et al.*

Skin wrinkling: can food make a difference? J. *Am. Coll. Nutr.*, 2001, 20: 71–80.

183. Pitt, J., Roth, W., Lacor, P. *et al.* Alzheimer's-associated Abeta oligomers show altered structure, immunoreactivity and synaptotoxicity with low doses of oleocanthal. *Toxicol. Appl. Pharmacol.*, 2009, 240: 189–197.

184. Beauchamp, G.K., Keast, R.S., Morel, D. *et al.* Phytochemistry: ibuprofen-like activity in extra-virgin olive oil. *Nature*, 2005, 437: 45–46.

185. Soffritti, M., Belpoggi, F., Manservigi, M. *et al.* Aspartame administered in feed, beginning prenatally through life span, induces cancers of the liver and lung in male Swiss mice. *Am. J. Ind. Med.*, 2010, 53: 1197–1206.

186. Wurtman, R.J. Neurochemical changes following high-dose aspartame with dietary carbohydrates. *N. Engl. J. Med.*, 1983, 309: 429–430.

187. Jacob, S.E. and Stechschulte, S. Formaldehyde, aspartame, and migraines: a possible connection. *Dermatitis*, 2008, 19: E10–E11.

188. Johns, D.R. Migraine provoked by aspartame. *N. Engl. J. Med.*, 1986, 315: 456.

189. Swithers, S.E., Davidson T.L. A role for sweet taste: calorie predictive relations in energy regulation by rats. *Behav. Neurosci.*, 2008 Feb; 122(1): 161–73.

190. Gregersen, S., Jeppesen, P.B., Holst, J.J. en Hermansen, K Antihyperglycemic effects of stevioside in type 2 diabetic subjects. *Metabolism*, 2004, 53: 73–76.

191. Stolarz-Skrzypek, K., Kuznetsova, T., Thijs, L. *et al.* Fatal and nonfatal outcomes, incidence of hypertension, and blood pressure changes in relation to urinary sodium excretion. JAMA, 2011, 305: 1777–1785.

192. Taylor, R.S., Ashton, K.E., Moxham, T., Hooper, L, en Ebrahim, S. Reduced dietary salt for the prevention of cardio-vascular disease: a meta-analysis of randomized controlled trials (Cochrane review). *Am. J. Hypertens.*, 2011, 24: 843–853.

193. Strazzullo, P., D'Elia, L., Kandala, N.B. en Cappuccio, F.P. Salt intake, stroke, and cardiovascular disease: meta-analysis of prospective studies. BMJ, 2009, 339: b4567.

194. D'Elia, L., Barba, G., Cappuccio, F.P. en Strazzullo, P. Potassium intake, stroke, and cardiovascular disease a meta-analysis of prospective studies. *J. Am. Coll. Cardiol.*, 2011, 57: 1210–1219.

195. Angiogenesis: An Integrative Approach from Science to Medicine. Edited by William D. Figg, M. Judah Folkman, Springer, 2008: 267.

196. Lamy, S., Bedard, V., Labbe, D. *et al.* The dietary flavones apigenin and luteolin impair smooth muscle cell migration and vegf expression through inhibition of PDGFR-beta phosphorylation. *Cancer Prev. Res.* (Phila), 2008, 1: 452–459.

197. Gupta S., Afaq F., Mukhtar H. Selective growth-inhibitory, cell-cycle deregulatory and apoptotic response of apigenin in normal versus human prostate carcinoma cells. *Biochem. Biophys. Res. Commun.*, 2001 Oct. 5; 287(4): 914–20.

198. Maggioni D., Garavello W., Rigolio R., Pignataro L. *et al.* Apigenin impairs oral squamous cell carcinoma growth in vitro inducing cell cycle arrest and apoptosis. *Int. J. Oncol.*, 2013 Nov; 43(5): 1675–82.

199. Ruef, J., Meshel, A.S., Hu, Z. *et al.* Flavopiridol inhibits smooth 194 cell proliferation in vitro and neointimal formation In vivo after carotid injury in the rat. *Circulation*, 1999, 100: 659–665.

200. Sekine, C., Sugihara, T., Miyake, S. *et al.* Successful treatment of animal models of reumatoid artritis with small-molecule cyclin-dependent kinase inhibitors. *J. Immunol.*, 2008, 180: 1954–1961.

201. Liu, J.Y., Lin, S.J. en Lin, J.K. Inhibitory effects of curcumin on protein kinase c activity induced by 12-O-tetradecanoyl-phorbol-13-acetate in NIH 3T3 cells. *Carcinogenesis*, 1993, 14: 857–861.

202. Korutla, L. en Kumar, R. Inhibitory effect of curcumin on epidermal growth factor receptor kinase activity in A431 cells. *Biochim. Biophys. Acta*, 1994, 1224: 597–600.

203. Hanif, R., Qiao, L., Shiff, S.J. and Rigas, B. Curcumin, a natural plant phenolic food additive, inhibits cell proliferation and induces cell cycle changes in colon adenocarcinoma cell

lines by a prostaglandin-independent pathway. *J. Lab Clin. Med.*, 1997, 130: 576–584.

204. Plummer, S.M., Holloway, K.A., Manson, M.M. *et al.* Inhibition of cyclo-oxygenase 2 expression in colon cells by the chemopreventive agent curcumin involves inhibition of NF-kappaB activation via the NIK/IKK signalling complex. *Oncogene*, 1999, 18: 6013–6020.

205. Lim, G.P., Chu, T., Yang, F. *et al.* The curry spice curcumin reduces oxidative damage and amyloid pathology in an Alzheimer transgenic mouse. *J. Neurosci.*, 2001, 21: 8370–8377.

206. Frautschy, S.A., Hu, W., Kim, P. *et al.* Phenolic anti-inflammatory antioxidant reversal of Abeta-induced cognitive deficits and neuropathology. *Neurobiol. Aging*, 2001, 22: 993–1005.

207. Bala, K., Tripathy, B.C. and Sharma, D. Neuroprotective and anti-ageing effects of curcumin in aged rat brain regions. *Biogerontology.*, 2006, 7: 81–89.

208. Shoba, G., Joy, D., Joseph, T. *et al.* Influence of piperine on the pharmacokinetics of curcumin in animals and human volunteers. *Planta Med.*, 1998, 64: 353–356.

209. Jang, S., Dilger, R.N. en Johnson, R.W. Luteolin inhibits microglia and alters hippocampal-dependent spatial working memory in aged mice. *J. Nutr.*, 2010, 140: 1892–1898.

210. Frydman-Marom, A., Levin, A., Faifara, D. *et al.* Orally administrated cinnamon extract reduces beta-amyloid oligo merization and corrects cognitive impairment in Alzheimer's disease animal models. *PLoS Biology*, 2011, 6: e16564.

211. Khan, A., Safdar, M., Ali Khan, M.M., Khattak, K.N., and Anderson, R.A. Cinnamon improves glucose and lipids of people with type-2-diabetes. *Diabetes Care*, 2003, 26: 3215–3218.

212. Dorant, E., van den Brandt, P.A., Goldbohm, R.A., and Sturmans, F. Consumption of onions and a reduced risk of stomach carcinoma. *Gastroenterology*, 1996, 110: 12–20.

213. Challier, B., Perarnau, J.M. and Viel, J.F. Garlic, onion and cereal fibre as protective factors for breast cancer: a French

case-control study. *Eur. J. Epidemiol.*, 1998, 14: 737–747.

214. Rahman, K. en Lowe, G.M. Garlic and cardiovascular disease: a critical review. *J. Nutr.*, 2006, 136: 736S-740S.

215. Ford, L., Graham, V., Wall, A. and Berg, J. Vitamin D concentrations in an UK inner-city multicultural outpatient population. *Ann. Clin. Biochem.*, 2006, 43: 468–473.

216. Garland, C.F., Comstock, G.W., Garland, F.C. *et al.* Serum 25-hydroxyvitamin D and colon cancer: eight-year prospective study. *Lancet*, 1989, 2: 1176–1178.

217. Malabanan, A., Veronikis, I.E. en Holick, M.F. Redefining vitamin D insufficiency. *Lancet*, 1998, 351: 805–806.

218. Davis, D.R., Epp, M.D. en Riordan, H.D. Changes in USDA food composition data for 43 garden crops, 1950 to 1999. *J. Am. Coll. Nutr.*, 2004, 23: 669–682.

219. Mayer, A.M. Historical changes in the mineral content of fruits and vegetables. *British Food Journal* 99 (6), 207–211. 1997.

220. Omenn, G.S. Chemoprevention of lung cancer: the rise and demise of beta-carotene. *Annu. Rev. Public Health*, 1998, 19: 73–99.

221. Podmore, I.D., Griffiths, H.R., Herbert, K.E. *et al.* Vitamin C exhibits pro-oxidant properties. *Nature*, 1998, 392: 559.

222. Leone, N., Courbon, D., Ducimetiere, P. en Zureik, M. Zinc, copper, and magnesium and risks for all-cause, cancer, and cardiovascular mortality. *Epidemiology*, 2006, 17: 308–314.

223. Rimm, E.B., Willett, W.C., Hu, F.B. *et al.* Folate and vitamin B6 from diet and supplements in relation to risk of coronary heart disease among women. JAMA, 1998, 279: 359–364.

224. Vogiatzoglou, A., Refsum, H., Johnston, C. *et al.* Vitamin B12 status and rate of brain volume loss in community-dwelling elderly. *Neurology*, 2008, 71: 826–832.

225. Gwenaelle Douaud *et. al.* Preventing Alzheimer's disease-related gray matter atraphy by B vitamin treatment. *Proceedings of the National Academy of Sciences*, 2013.

226. Armitage, J.M., Bowman, L., Clarke, R.J. *et al.* Effects of homocysteine-lowering with folic acid plus vitamin B12 vs placebo on mortality and major morbidity in myocardial

infarction survivors: a randomized trial. JAMA, 2010, 303: 2486–2494.

227. Schnyder, G., Roffi, M., Flammer, Y., Pin, R., and Hess, O. M. Effect of homocysteine-lowering therapy with folic acid, vitamin B12, and vitamin B6 on clinical outcome after percutaneous coronary intervention: the Swiss Heart study: a randomized controlled trial. jama, 2002, 288: 973–979.

228. Verheesen, R.H. and C.M. Schweitze. Het jodiumtekort is terug. *Medisch Contact*, 24–10–2008.

229. Andersson, M., *et al.* Iodine deficiency in Europe: a continuing public health problem. World Health Organization, 2007.

230. Baum *et al.*High risk of hiv-related mortality is associated with selenium deficiency. *J Acquir Immune Defic Syndr Hum Retrovirol.* 1997 Aug 15; 15 (5): 370–4.

231. Knekt, P., Heliovaara, M., Aho, K. *et al.* Serum selenium, serum alpha-tocopherol, and the risk of reumatoid artritis. *Epidemiology*, 2000, 11: 402–405.

232. Yoshizawa, K., Willett, W.C., Morris, S.J. *et al.* Study of prediagnostic selenium level in toenails and the risk of advanced prostate cancer. *J. Natl. Cancer Inst.*, 1998, 90: 1219–1224.

233. Clark, L.C., Combs jr., G.F., Turnbull, B.W. *et al.* Effects of selenium supplementation for cancer prevention in patients with carcinoma of the skin. A randomized controlled trial. Nutritional Prevention of Cancer Study Group. JAMA, 1996, 276: 1957–1963.

234. Macpherson, A. et al. Loss of Canadian wheat lowers selenium intake and status of the Scottish populuation. National Research Council of Canada, 1997.

235. Meltzer, H.M., et al. Different bioavailability in humans of wheat and fish selenium as measured by blood platelet response to increased dietary selenium. *Biological Trace Element Research*, 1992, 36: 229–241.

236. Gissel T., Rejnmark L., Mosekilde L., Vestergaard P.J. Intake of vitamin D and risk of breast cancer: a meta-analysis. *Steroid Biochem Mol Biol*, 2008 Sep; 111(3–5): 195–9.

237. Autier P., Gandini S. Vitamin D supplementation and total mortality: a meta-analysis of randomized controlled trials. *Arch Intern Med*, 2007 Sep 10; 167(16): 1730–7.

238. Tavera-Mendoza, L.E., en John H.White. Cell defenses and the sunshine vitamin. *Scientific American*, 2008.

239. Burton, J.M., Kimball, S., Vieth, R. *et al.* A phase I/II dose-escalation trial of vitamin d3 and calcium in multiple sclerosis. *Neurology*, 2010, 74: 1852–1859.

240. Stalpers-Konijnenburg, S.C., et al. Waar is de zon die mij zal verblij-den . . .': vitamine D deficiëntie en depressie bij ouderen. *Tijdschrift voor psychiatrie*, 2011.

241. Scragg, R., Jackson, R., Holdaway, I.M., Lim, T. en Beaglehole, R. Myocardial infarction is inversely associated with plasma 25-hydroxyvitamin D3 levels: a community-based study. *Int. J. Epidemiol.*, 1990, 19: 559–563.

242. Vieth, R. Vitamin D supplementation, 25-hydroxyvitamin D concentrations, and safety. *Am. J. Clin. Nutr.*, 1999, 69: 842–856.

243. *Atlas of Multiple Sclerosis.* World Health Organization, 2008.

244. Yang, Y.X., Lewis, J.D., Epstein, S. en Metz, D.C. Long-term proton pump inhibitor therapy and risk of hip fracture. *JAMA*, 2006, 296: 2947–2953.

245. Wenner Moyer, M., Heartburn Headache: Overuse of Acid Blockers Poses Health Risks. *Scientific American*, 2010.

246. Gulmez, S.E., Holm, A., Frederiksen, H. *et al.* Use of proton pump inhibitors and the risk of community-acquired pneumonia: a population-based case-control study. *Arch. Intern. Med.*, 2007, 167: 950–955.

247. Howell, M.D., Novack, V., Grgurich, P. *et al.* Iatrogenic gastric acid suppression and the risk of nosocomial Clostridium difficile infection. *Arch. Intern. Med.*, 2010, 170: 784–790.

248. Michielsen, P. *Toxische leverbeschadiging, cursus leverziekten.* Universiteit Antwerpen, 2011.

249. Lee, I.M., Djousse, L., Sesso, H.D., Wang, L. and Buring, J.E. Physical activity and weight gain prevention. *JAMA*, 2010, 303: 1173–1179.

250. Rovio, S., Kareholt, I., Helkala, E.L. *et al.* Leisure-time physical activity at midlife and the risk of dementia and Alzheimer's disease. *Lancet Neurol.*, 2005, 4: 705–711.

251. Willey, J.Z., Moon, Y.P., Paik, M.C. *et al.* Lower prevalence of silent brain infarcts in the physically active: the Northern Manhattan Study. *Neurology*, 2011, 76: 2112–2118.

252. Craft, L.L. and Perna, F.M. The Benefits of Exercise for the Clinically Depressed. Prim. Care Companion. *J. Clin. Psychiatry*, 2004, 6: 104–111.

253. Erickson, K.I., Voss, M.W., Prakash, R.S. *et al.* Exercise training increases size of hippocampus and improves memory. *Proc. Natl. Acad. Sci. U.S.A*, 2011, 108: 3017–3022.

254. Walford, Roy. *Beyond the 120 year diet.* 2000, Thunder's Mouth Press, New York.

255. Handschin, C. and Spiegelman, B.M. The role of exercise and PG-C1alpha in inflammation and chronic disease. *Nature*, 2008, 454: 463–469.

256. Adams, S.A., Matthews, C.E., Hebert, J.R. *et al.* Association of physical activity with hormone receptor status: the Shanghai Breast Cancer Study. Cancer Epidemiol. *Biomarkers Prev.*, 2006, 15: 1170–1178.

257. Giovannucci, E.L., Liu, Y., Leitzmann, M.F., Stampfer, M.J. and Willett, W.C.A prospective study of physical activity and incident and fatal prostate cancer. *Arch. Intern. Med.*, 2005, 165: 1005–1010.

258. Servan-Schreiber, D., *Antikanker.* 2006, Kosmos Uitgevers.

259. Walford, Roy. *Beyond the 120 year diet.* 2000, Thunder's Mouth Press, New York.

260. Colman, R.J., Anderson, R.M., Johnson, S.C. *et al.* Caloric restriction delays disease onset and mortality in resus monkeys. *Science*, 2009, 325: 201–204.

261. Mattison, J.A., Roth G.S., Beasley T.M., Tilmont E.M. *et al.* Impact of caloric restriction on health and survival in rhesus monkeys from the NIA study. *Nature*, 2012 Sep 13; 489(7415): 31821.

262. Vallejo, E.A. Hunger diet on alternate days in the nutrition of the aged. *Prensa. Med. Argent*, 1957, 44: 119–120.

263. *The Fast Diet*, dr. Michael Mosley and Mimi Spencer, Atria books, New York, 2013.

264. Duncan, K.H., Bacon, J.A. and Weinsier, R.L. The effects of high and low energy density diets on satiety, energy intake,

and eating time of obese and nonobese subjects. *Am. J. Clin. Nutr.*, 1983, 37: 763–767.

265. Whitmer, R.A., Gustafson, D.R., Barrett-Connor, E. *et al.* Central obesity and increased risk of dementia more than three decades later. *Neurology*, 2008, 71: 1057–1064.

266. Kawachi, I., Sparrow, D., Kubzansky, L.D. *et al.* Prospective study of a self-report type A scale and risk of coronary heart disease: test of the MMPI-2 type a scale. *Circulation*, 1998, 98: 405–412.

267. Hjemdahl, P., Annika Rosengren Andrew Steptoe *et al. Stress and Cardiovascular Disease.* Springer, 2011.

268. Orth-Gomer, K. and Leineweber, C. Multiple stressors and coronary disease in women. The Stockholm Female Coronary Risk Study. *Biol. Psychol.*, 2005, 69: 57–66.

269. Kooy, K. van der, Marwijk, H. van *et al.* Depression and the risk for cardiovascular diseases: systematic review and meta analysis. *Int. J. Geriatr. Psychiatry*, 2007, 22: 613–626.

270. Visintainer, M.A., Volpicelli, J.R. and Seligman, M.E. Tumor rejection in rats after inescapable or escapable shock. *Science*, 1982, 216: 437–439.

271 Schneider, R., Nidich, S., Morley Kotchen, J. *et al.* Effects of Stress Reduction on Clinical Events in African Americans With Coronary Heart Disease: A Randomized Controlled Trial. *Circulation*, 2009, 120: S461.

272. Lazar, S.W., Kerr, C.E., Wasserman, R.H. *et al.* Meditation experience is associated with increased cortical thickness. *Neuroreport*, 2005, 16: 1893–1897.

273. Previc, F.H. The role of the extrapersonal brain systems in religious activity. *Conscious. Cogn*, 2006, 15: 500–539.

274. Tsuji, H., Venditti jr., F.J., Manders, E.S. *et al.* Reduced heart rate variability and mortality risk in an elderly cohort. The Framingham Heart Study. *Circulation*, 1994, 90: 878–883.

275. Rodin, J. and Langer, E.J. Long-term effects of a control-relevant intervention with the institutionalized aged. *J. Pers. Soc. Psychol.*, 1977, 35: 897–902.

276. Friedmann, E. and Thomas, S.A. Pet ownership, social support, and one-year survival after acute myocardial infarction in the

Cardiac Arrhythmia Suppression Trial (cast). *Am. J. Cardiol.*, 1995, 76: 1213–1217.

277. Spiegel, D., Bloom, J.R., Kraemer, H.C. and Gottheil, E. Effect of psychosocial treatment on survival of patients with metastatic breast cancer. *Lancet*, 1989, 2: 888–891.

278. Spiegel, D., Butler, L.D., Giese-Davis, J. *et al.* Effects of supportive-expressive group therapy on survival of patients with metastatic breast cancer: a randomized prospective trial. *Cancer*, 2007, 110: 1130–1138.

279. Weindruch, R. and Walford, R.L. Dietary restriction in mice beginning at 1 year of age: effect on life-span and spontaneous cancer incidence. *Science*, 1982, 215: 1415–1418.

280. Walford, Roy. *Beyond the 120 year diet.* 2000, Thunder's Mouth Press, New York.

281. Walford, R.L., Mock, D., MacCallum, T. and Laseter, J.L. Physiologic changes in humans subjected to severe, selective calorie restriction for two years in biosphere 2: health, aging, and toxicological perspectives. *Toxicol. Sci.*, 1999, 52: 61–65.

282. Rose, M.R. *The Long Tomorrow: How Advances in Evolutionary Biology Can Help Us Postpone Aging.* 2005, Oxford University Press, usa.

283. Brandt, P.A., van den. The impact of a Mediterranean diet and healthy lifestyle on premature mortality in men and women. *Am. J. Clin. Nutr.*, 2011, 94: 913–920.

Index

butter 36, 38, 43, 93–95, 194, 214–216, 301
button mushrooms 158, 164

cabbage 148, 154, 168, 171, 315–316, 332
calcium 38, 143–146, 148, 191, 239, 252, 259–262
calcium supplements 36, 142
calorie restriction 29, 274–278, 280–281, 283–284, 305
   calorie restriction under optimal nutrition (CRON) 29, 280
camomile 232
Canada 108, 257
cancer 8, 16–17, 19, 23–25, 27, 30, 32, 49, 57, 60–65, 78, 80–81, 84, 115, 125, 127–132, 134, 138, 141, 143–144, 147–148, 160, 162–166, 168, 171–174, 182, 184, 186–187, 190, 195, 197, 206, 208–209, 219, 225, 227–231, 233–235, 237–238, 240–241, 243–244, 247–248, 250–254, 266, 271–273, 275–277, 283, 287, 295, 340, 351
   cancer cells 61–64, 225–228, 230, 233, 243–244, 250, 254, 287
canola oil 217–219
capers 234, 320–321
carbohydrates 49–51, 54–55, 65, 69, 71, 79, 85, 164, 171, 201, 240, 301–302, 342–343, 346, 350
carbon atoms 51–53, 72–73, 87–90, 92, 96–97, 349
carcinogens 220
cardamom 310
cardiovascular diseases 22, 24–27, 32, 59–60, 84, 86–87, 94, 97, 99–100, 102, 105, 108, 117, 120, 161, 167, 206, 208, 210,

215, 218, 223–225, 234–235, 240–242, 247, 249, 252, 255–258, 262, 270, 275–276, 284, 286, 293, 304, 339–342, 346
   see also vascular disease
cardiovascular system 86, 204, 234, 241, 338, 343
   see also vascular system
carnosine 82, 84, 190
carotene 127, 178, 240–242
carrots 154, 171, 173, 240, 312, 315–316
catalases 249
cataracts 9, 59, 64, 167
cauliflower 171, 174, 202, 204
cayenne 316, 326
celery 173, 315–316, 329
cell check proteins 230–231
cell walls 104, 106–107, 113, 126, 346
cereals 34, 152–153, 159, 340
cerebral cortex 289, 347
cerebral infarctions 268
cervical cancer 130
cheese 38, 43–44, 46, 95, 120, 148, 190, 194–196, 202, 204, 321, 324, 327–328, 330–331, 341
chemotherapy 15, 18, 165, 227, 244, 260
cherries 154
chicken 38, 44, 171, 190, 192–194, 322, 326–327, 341
chickpeas 321–322
chicory 171
chives 330–332
chlorine atoms 263
chocolate 43, 69, 95, 120, 205, 267, 342
   see also dark chocolate
cholesterol 26–27, 33, 60, 85–86, 103–106, 194–196, 213–216, 218–219, 260, 279, 340, 351

free-range chickens 192
fried food 43, 94–95, 119, 168, 197–198, 301
fructose 51–55, 154, 222
fruit juice 42–44, 69, 121, 132, 151, 171, 177–180, 189, 209, 211, 214, 308, 311, 313–314, 317–320, 324, 326, 341
  shop-bought fruit juices 132, 134, 148
fruits 34–38, 41–43, 69, 79, 146, 149–151, 166–169, 172, 174, 176–178, 180, 182, 187–189, 195–196, 202–204, 207, 209, 212, 214, 218–219, 222, 224–225, 229, 238, 240, 259, 266, 281–282, 296, 301, 303, 305, 307–308, 310–311, 316, 333, 339–341
  blue fruits 180, 189, 314
  dried fruit 157–158, 202
  fruit sugar 52, 187–188
  red fruits 180, 189, 314
Furhman diet 29–30

gallbladder disorders 102
gallstones 136
garlic 42, 154, 172–173, 214, 234–235, 308, 316–318, 321–323, 327, 330–332
gastric acid 260–262, 264
  gastric acid inhibitors 260, 266
gastroenterology 135, 146, 292
genital warts 129–130
germ 155–156
Germany 108
ginger 323
ginger cookie spices 310
ginger tea 43, 46, 121, 125, 131–132, 147–148, 313
gingerol 131
Gleevec 18, 226–228
glucose 51–56, 58, 60, 63–64,

132, 153–156, 158, 164, 197, 199, 215, 221–222, 279, 340, 350
glucose level 55, 152–153, 159, 162
glutamic acid 73
glutamine 73
glutathione 234
glutathione peroxidase 249–251
gluten 321
  gluten sensitivity 80, 84, 262
glycaemic index 65, 85–86, 153–158, 188, 321, 339–340, 342–343, 350
glycaemic load 154–155, 339, 342–343
glycerol 87
glycine 71, 73
glycolipids 106
glycolysis 62–63
goat cheese 324, 327
grain 36–37, 39, 70, 150, 155–156, 212, 251, 321
  grain industry 37, 47, 166
grapefruit 154, 173, 309
grapes 151, 158, 202–203, 296, 308–309, 312, 314, 333
  grape sugar 56
green tea 11, 22, 30, 43–44, 46–47, 121, 125–132, 135, 137, 147–148, 151, 164, 168, 173, 189, 196, 204, 208, 284, 308–309, 313
growth 55–57, 61, 64, 67, 77, 81, 182, 274, 278–279, 350
  growth factors 144, 226–227

hamburgers 43, 190, 197, 212
Harvard 13, 36, 38–39, 80, 142, 154–155, 170, 239, 250, 267, 339
Hawaii 161
hazelnuts 201
  hazelnut oil 324

Acknowledgements

I would like to thank Professor Stephen Spindler of the University of California (Riverside) and Dr Aubrey de Grey of Cambridge University for our interesting discussions about the ageing process and nutrition. Closer to home, I would like to thank Dr Herman Becq, Dr Tania Daems, Dr Hans Decoster, Sven Bulterijs and Adjiedj Bakas for their insights and advice. Also thanks to Paloma Sánchez van Dijck of Uitgeverij Prometheus/Bert Bakker and also to many more people; instructors, professors, physicians and patients who have given me valuable insights, not only in terms of science, but also in terms of people.

www.foodhourglass.com

Join the Facebook page of Kris Verburgh for regular updates about health, ageing and nutrition.

Via the site *foodhourglass.com* you can:

- subscribe to the newsletter to stay informed on news about the food hourglass, nutrition and ageing;
- assess your 'real' age via a questionnaire developed by Kris Verburgh;
- download an image of the food hourglass in colour;
- discover healthy new recipes;
- find much more extra information.